DOUG SCOTT
Kangchen-junga
THE HIMALAYAN GIANT

DOUG SCOTT
Kangchenjunga

THE HIMALAYAN GIANT

Revised & edited by
Catherine Moorehead

Vertebrate Publishing, Sheffield
www.v-publishing.co.uk

DOUG SCOTT
Kangchenjunga

Revised and edited by Catherine Moorehead.
First published in 2021 by Vertebrate Publishing.

Vertebrate Publishing
Omega Court, 352 Cemetery Road, Sheffield S11 8FT, United Kingdom.
www.v-publishing.co.uk

Copyright © Doug Scott 2021.

Front cover: Approaching the summit of Kangchenjunga, 1979. © Doug Scott.
Author photo © Chris Bonington.
Photography by Doug Scott unless otherwise credited.

Mapping contains Openstreetmap.org data © OpenStreetMap contributors, licence www.openstreetmap.org/copyright, with relief shading and contours produced from data derived from U.S. Geological Survey, National Geospatial Program.
Cartography by Richard Ross, Active Maps Ltd – www.activemaps.co.uk

Doug Scott has asserted his rights under the Copyright, Designs and Patents Act 1988 to be identified as author of this work.

This book is a work of non-fiction. The author has stated to the publishers that, except in such minor respects not affecting the substantial accuracy of the work, the contents of the book are true.

A CIP catalogue record for this book is available from the British Library.

ISBN: 978-1-912560-19-6 (Hardback)
ISBN: 978-1-912560-20-2 (Ebook)
ISBN: 978-1-83981-074-9 (Audiobook)

10 9 8 7 6 5 4 3 2 1

All rights reserved. No part of this work covered by the copyright herein may be reproduced or used in any form or by any means – graphic, electronic, or mechanised, including photocopying, recording, taping or information storage and retrieval systems – without the written permission of the publisher.

Every effort has been made to obtain the necessary permissions with reference to copyright material, both illustrative and quoted. We apologise for any omissions in this respect and will be pleased to make the appropriate acknowledgements in any future edition.

Cover design by Jane Beagley, Vertebrate Publishing.
Production by Cameron Bonser, Vertebrate Publishing.
www.v-publishing.co.uk

Vertebrate Publishing is committed to printing on paper from sustainable sources.

Printed and bound in the UK by TJ Books Limited, Padstow, Cornwall.

Contents

Introduction		VII

PART 1   **The Mountain and Its People**
1	The Kangchenjunga Massif	5
2	The Peoples of Eastern Nepal and Sikkim	17

PART 2   **Western Approaches**
3	Missionaries, Traders and Politicians	35
4	Measuring the Heights	47
5	The Expeditions of J.D. Hooker	63
6	Artists, Writers and Photographers	83

PART 3   **Climbs and Attempts**
7	The Pioneers	103
8	The Kabru Controversy	123
9	Crowley and the Norwegians	135
10	Kellas, Raeburn and Crawford	151
11	Between the Wars I: British and American Expeditions	163
12	Between the Wars II: the 1930 International Himalayan Expedition	181
13	Between the Wars III: the German Expeditions	189
14	War, Partition and Reconnaissance	207

PART 4   **Ascents**
15	The First and Second Ascents	221
16	The First Ascent without Supplementary Oxygen	233

The Kangchenjunga Massif: Treks, Attempts and Ascents	259
Select Bibliography	261
Index	267

Introduction

A promontory of high ground, the Singalila Ridge, aligned north–south to the main Himalayan range, divides the Tista river basin in Sikkim from that of the Tamur in Nepal. This subsidiary series of mountains and hills extends for 160 kilometres through the foothills of the Himalaya to eventually merge with the northern plains of the Indian sub-continent. The highest point of this chain of mountains is Kangchenjunga, the third-highest of the world's mountains.

From the east, Kangchenjunga and all its adjacent peaks appear as a great wall of rock, ice and snow, to present a formidable barrier that both divides and protects the peoples of Sikkim and Eastern Nepal.

From the flanks of Makalu in the west, Kangchenjunga is instantly recognisable, as it was for me from Everest in 1975 and again from Makalu in 1980, on the far edge of a massive cloud inversion completely filling the intervening Arun river basin. The sea of white mist lapped against the peninsulas and jutting spurs, invading all the intervening bays to the eastern shoreline against the flanks of the Singalila Ridge. This magical scene lasted a couple of days until it was lost, smothered in cloud for the next week of wild, stormy weather.

Kangchenjunga has huge religious significance, second only to the most holy mountain, Kailash. The reclusive Lepcha people of Sikkim, living in the shadow of Kangchenjunga, according to Laurence Waddell in his *Among the Himalayas* (1899), long ago gave to Kangchenjunga the name of *Kong-Lo-Chu* or 'The highest screen or curtain of the snows'. In the excellent book *Khangchendzonga: Sacred Summit* (2007), Pema Wangchuk writes that Kangchenjunga is also known to the Lepcha as *Kingtsoom Zaongboo Chu*, 'The auspicious forehead peak', the highest veil of snow beyond which the spirits of their ancestors dwell in Rum Lyang, the land of the gods.

It was in the eighth century, when bringing Buddhism to the Himalayan kingdoms, that Padmasambhava (Guru Rinpoche) declared Sikkim to be the most sacred of all the *beyul* (hidden lands) and Kangchenjunga the most important of all the guardian deities. The Bhutia, coming from Tibet in the sixteenth century, recognised the mountain as sacred and gave it the name that has now become known today as Kangchenjunga (with many variations). There are many interpretations of what exactly the name means. The commonly accepted explanation is 'Five treasures of great snow', which are said to correspond to the five summits of the mountain. The treasures are thought to be holy texts, grains, gold or turquoise, salt and weapons. There are several opinions about the above, including one by the artist and mystic Nicholas Roerich in *Himalayas: Abode of Light* (1947). He writes:

> All eyes are attracted to the majestic white summits beyond the clouds, as if rising over an inferior world. From all lands, the highest hopes are directed to the Himalayas. Kang-chen-zod-nga – Five Treasures of the Great Snow. And why is this gorgeous mountain so called? It is because it contains a store of the five most precious things in the world. What things? – gold, diamonds and rubies? By no means. The ancient East values other treasures. It is said that there will come a time when famine will overtake the whole world. At that time a Man will appear who will unlock the giant gate of these vast treasuries and will nourish all mankind. And it is of course understood that this Man will nourish humanity not with material food, but with spiritual food.

The first foreigners to explore Kangchenjunga were the British, who were in a privileged position for access to the mountain. They set the scene, providing details of the terrain, producing maps, building infrastructure and generally disseminating information on the climate, vegetation and indigenous population.

During the latter half of the nineteenth century, groups of visitors arrived to explore the valleys radiating out from Kangchenjunga, slowly at first – just one or two a year – and then, in the twentieth century, in greater numbers; until now, in these days of mass tourism, they threaten the peace and solitude the visitor pays to experience, and harm both the local culture and the environment. Included in this are the mountaineers who go on to the higher reaches of the mountain. In recent years, however, mountaineers have tended to arrive in smaller groups climbing in lightweight style,

INTRODUCTION

leaving no trace and having paid attention to the health and wellbeing of the local people employed to make their climbs possible.

The exploration of the upper part of the mountain naturally comes at the end of this book. First the reader is taken back to the beginning of this voyage of discovery – literally from sea to summit, geologically and historically, from what the scientists have revealed and from what the historians have gleaned of man's advance across northern India in prehistoric times towards the Eastern Himalaya and the environs of Kangchenjunga.

PART 1
The Mountain and Its People

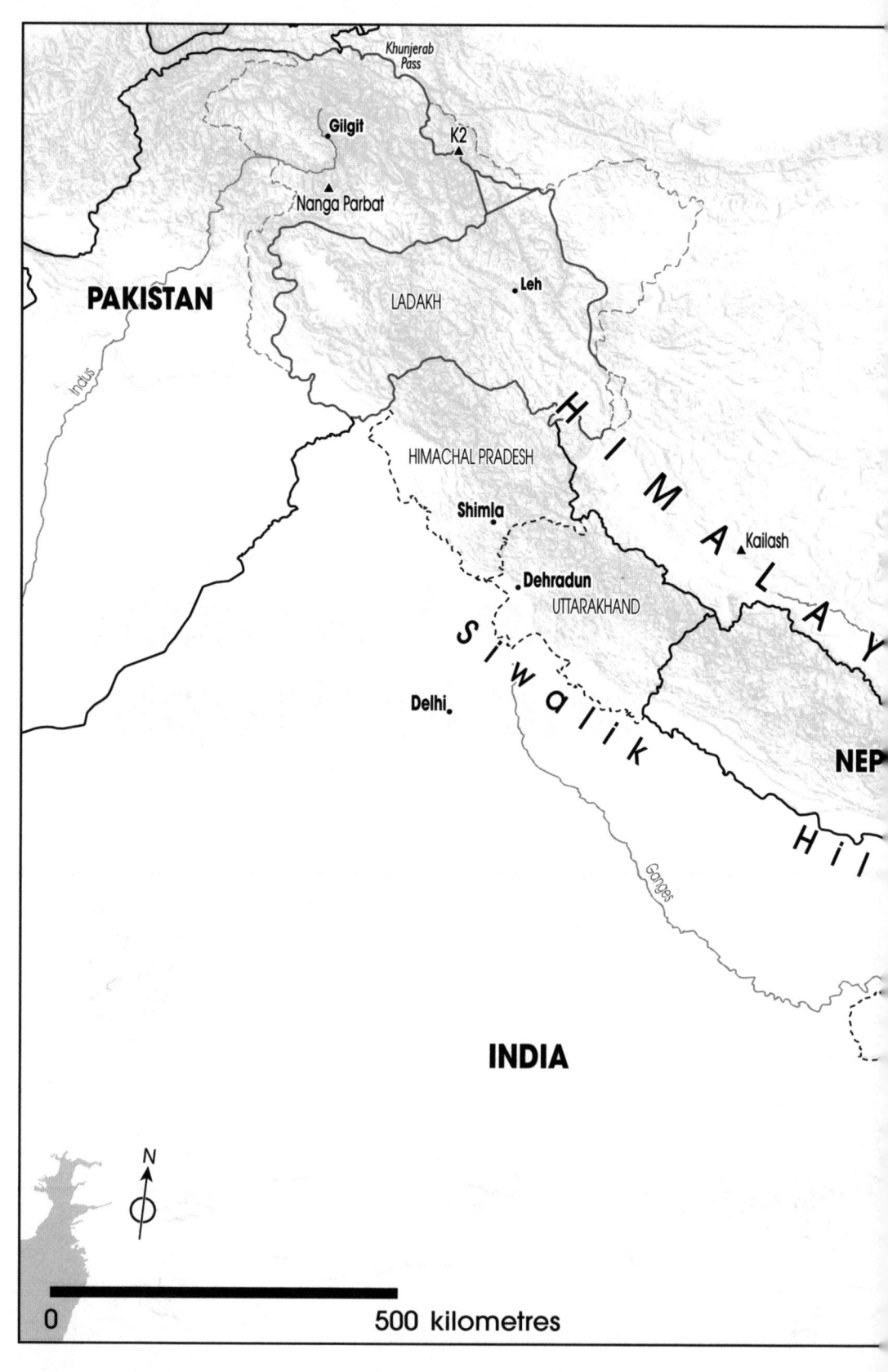

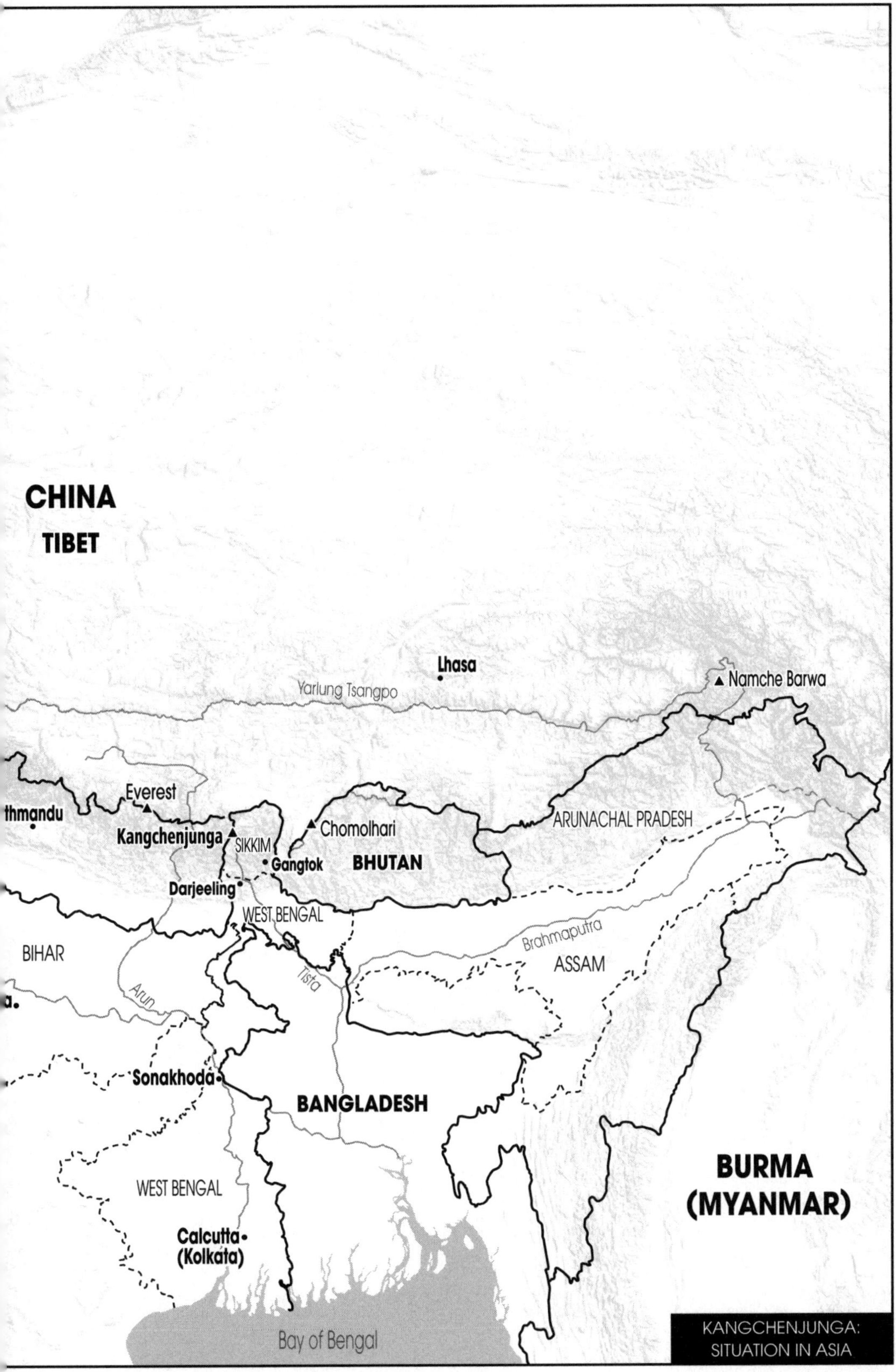
CHINA
TIBET
Lhasa
Yarlung Tsangpo
Namche Barwa
Everest
thmandu
Kangchenjunga
SIKKIM
Chomolhari
Gangtok
BHUTAN
ARUNACHAL PRADESH
Darjeeling
WEST BENGAL
BIHAR
Brahmaputra
Tista
ASSAM
Arun
Sonakhoda
BANGLADESH
WEST BENGAL
BURMA
(MYANMAR)
Calcutta
(Kolkata)
Bay of Bengal
KANGCHENJUNGA:
SITUATION IN ASIA

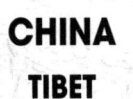

1

The Kangchenjunga Massif

The Himalaya, about 2,500 kilometres in length, is not the most extensive mountain chain, but it has proved to be the most significant in influencing human history, largely because of having arisen between two continents, whereas the Andes, for example, are formed along the edge of a continent far from the world's heartlands.

The people who lived in sight of the Himalaya in ancient times could hardly fail to be impressed by these snow-capped peaks:

> In a hundred ages of the gods I could not describe to you the glories of Himachal … As the dew is dried up by the morning sun, so are the sins of mankind by the sight of Himachal.[1]

Our forebears, living lives governed by natural processes, would no doubt have been profoundly affected whenever a glimpse of the Himalaya burst upon their senses. Even today, we, the beneficiaries and victims of modern civilisation, may still have our spirits raised by the sight of the Himalaya.

The earliest reference to the Himalaya comes in the Vedic peoples' Sanskrit texts, first written during the Bronze Age, around 1500 BC. A verse of the *Rig Veda* states:

> I will declare the mighty deeds of Vishnu, of Him who measured out the earthly regions … Let the hymn lift itself as strength to Vishnu, the Bull far striding, dwelling on the mountains …[2]

---

1. Quoted from the *Puranas* (ancient Hindu mythological writings) in *Sacred Mountains of the World*, Edwin Bernbaum, University of California Press, Berkeley, California, 1998.
2. *Rig Veda: Hymn CLIV*, translated by Ralph T.H. Griffith, 1896, published at *www.sacredtexts.com*

The Vedic texts, especially the longest poem, the *Mahabharata*, written in the fourth century BC, discuss human spirituality. The *Mahabharata's* 200,000 verses refer to many mountain peoples, such as the Pavatas, who lived in what is now Nepal, the kingdom of Nepa, or the Kiranti kingdom, now encompassing the Eastern Himalaya, to which Arjuna, this epic's main protagonist, made many expeditions.

When these tales were first being told, mankind was experiencing the effect of inter-glacial warming. A huge volume of water was released at the end of the last Ice Age, when glacier ice was connected all the way from Kangchenjunga to Nanga Parbat and across the Karakoram to the ice sheets of the Tibetan plateau. The tales reflect the dramatic changes taking place as the ice melted: sea levels rose while land became unstable; fires burned and global warming prevailed. All this was passed through the generations by a strict oral tradition first written down around 1500 BC – and not just in India.

Geological origins

Around 175 million years ago, the ancient super-continent of Pangaea gradually broke into separate, drifting 'plates'.

The Indian Plate gradually moved northwards at about twenty centimetres a year across the Indian Ocean. Approximately fifty million years ago and roughly near the equator, this plate collided with Asia, closing over the Tethys Ocean that once lay between them. This great continental collision resulted in the uplift of the Himalaya along the leading margin of India and the plateau of Tibet, the largest and highest continental clash of its kind on Earth. Within the crust, rocks metamorphosed as temperatures and pressures increased with depth. Through the middle part of the crust, temperatures became so high that the rocks started to melt. These fused rocks, or 'migmatites', led to the formation of granites. The Kangchenjunga massif is made up entirely of these migmatites and leuco- or light-coloured granites. Sheets of these can be seen along Kangchenjunga's south face, where thin black lines of rock mark these sill margins where the magma was squeezed out from its melting source beneath the Tibetan plateau and so to the surface along the Great Himalayan range. These Kangchenjunga leuco-granites contain beautiful small crystals of red garnet, black tourmaline and white mica. (The Indian Plate continued relentlessly to push north and lifted the Himalayan summits, including Kangchenjunga, even higher, at a present estimated rate of about one and a third centimetres per annum.)

The three mountain ranges

Mountain travellers in Vedic times had returned to confirm that the Himalaya consisted of three parallel ranges, labelled 'Inner Mountains', 'Outer Mountains' and 'Small Mountains'. These were the first known accounts of trying to classify the geography of the Himalaya, yet they correspond with recent observations.

The Siwaliks ('Small Mountains')

The Siwaliks are the southernmost mountains of the system and are sometimes known as the 'Outer Himalaya'.[3] They stretch from the Indus almost to Assam. They vary in width from ten to fifty kilometres and from 1,500m to 2,000m in height. A gap in this range of about 300 kilometres occurs east of the Kosi river of East Nepal. The strength of the monsoon in this region has led to drastic erosion, especially from the Tista river of Sikkim, where only remnants of the Siwaliks remain.

This river erosion has helped to reveal rich fossil sites of large prehistoric animals. The first person to discover these in the Siwaliks was the Scottish geologist and botanist Hugh Falconer (1808–1865), from Forres.[4]

These hills' appearance naturally reflects the underlying geology, made up of tertiary deposits brought down by ice sheets and washed down by rivers from the high Himalaya. Many sub-ranges of mainly hump-backed ridges, bisected by large rivers running from north to south through Nepal to the Terai, characterise the chain.

The Siwaliks generally show a steep southern side dropping to the Terai and the flat northern edge of the Gangetic plain. While teeming with wildlife, it supports a scattered population owing to the rugged terrain and the dense, damp forests where malaria is virulent.

The Mahabharata Range ('Outer Mountains')

This significant east–west range rises to between 3,700m and 4,500m. These mountains are sometimes labelled the 'Lesser Himalaya' or 'Middle Himalaya'. They follow the same line as the Siwaliks but are much higher as well as older and more complex, averaging about 100 kilometres in width. Their southern slopes are generally steep, with a low population density, whereas on the gently sloping northern side, villages surrounded by terraced fields and lush pasture are relatively numerous.

These mountains have become a cultural and linguistic buffer zone

---

3   Not to be confused with 'Outer Mountains', as the Mahabharata Range is sometimes called.
4   The Falconer Museum in Forres, Moray, contains some of his finds and much else of interest.

between the highly populated plains of the south and the ever-growing population of the 'mid-lands'. They have also afforded the 'mid-lands' protection from the warring states of the Gangetic plain. Their northern slopes gradually merge with the Kathmandu Valley, with its dense population, well-preserved and wonderful Newari temples and palaces.[5] In this range, outside Nepal, the British established their hill stations such as Murree, Simla and Darjeeling. These have all expanded to become the home of thousands of residents and tourists escaping the searing heat of the Indian plains during April and May.

The Great Himalaya ('Inner Mountains')

This range includes ten of the world's fourteen 8,000m peaks. The crest of the main divide seldom drops below 5,000m except where the great rivers of High Asia have broken through. In doing so, they have divided the Himalaya into separate blocs which were first named by Sir Sidney Burrard in *A Sketch of the Geography and Geology of the Himalaya Mountains and Tibet* (1908) and extended by Kenneth Mason, the military surveyor and geographer, in his authoritative *Abode of Snow* (1955). Of these zones, the Nepal Himalaya contains along its northern border the most impressive concentration of snow-clad peaks anywhere.

The Nepal Himalaya is sub-divided into three sections from west to east: Karnali, Gandaki and Kosi; these sections are further sub-divided into Himals. East of the Arun lie the Kumbhakarna Himal and the Kangchenjunga Himal on the border with Sikkim. These two areas are the focus of this book.

The river systems

The Singalila Ridge, running south from Kangchenjunga, not only divides the Indian state of Sikkim from Eastern Nepal but also separates the drainage systems of both areas. This ridge marks the watershed between the Tista river flowing into the Brahmaputra, and the rivers of the Tamur–Arun–Kosi basin flowing into the Ganges. These systems coalesce in the delta area of the Bay of Bengal, into which they disgorge huge amounts of sediment.

All the glaciers on the west side of Kangchenjunga drain into the Tamur. In the spring, the waters are sparkling, clear and full of trout. From July to September, the monsoon rains cause them to turn into turgid brown torrents in full flood as they race down to join the Kosi. They erode the mountainsides deeply.

---

5   See Chapter 2, page 24.

On the eastern side of Kangchenjunga and the Singalila Ridge, all the land is drained by the Tista river and its many tributaries. In doing so, these have formed a huge basin which contains virtually all of Sikkim. The Tista, after traversing the whole country, flows south to cut through the heavily denuded sub-Himalayan Mahabharata Range between Darjeeling and Kalimpong.

The topography of the Kangchenjunga massif

Kangchenjunga (8,586m) is situated on the border between western Sikkim and eastern Nepal. It lies at the centre of a considerable massif with many inspiring and formidable satellite peaks. All possess degrees of religious significance for the local people. Unlike other great mountain massifs, such as Everest and Annapurna, Kangchenjunga is, from all directions, clearly a dominant summit. Of all the great mountains, Kangchenjunga is the most impressive: it is the 8,000m summit with the greatest religious significance.

This massif is demarcated by two passes: the Jonsong La in the north, and the Kang La to the south. Along the fifty kilometres of intervening ridge lie nine very prominent peaks, all worthy of attention from the traveller and Himalayan mountaineer. The most northerly peak is Langpo (6,965m).[6] To its south lies Pathibhara (The Pyramid, 7,164m), Kirat Chuli (Tent Peak, 7,365m), Nepal Peak (7,177m), Gimmigela I (7,350m) and then Gimmigela II (7,007m; together The Twins). South of the central massif, still on the main divide, stands Talung Peak (7,349m), Kabru North (c.7,338m), Kabru South (7,318m), Rathong Peak (6,682m) and Kokthang (6,147m).[7] Well to the south of this stunning group comes the Kang La (c.5,054m), the massif's southern limit.

The central mass includes Kangchenjunga South Summit (8,494m). Between the Main Summit and the South Summit, a Central Summit has been identified and given the height of 8,496m. I have to say that this peak was hardly noticeable when I stood on the Main Summit looking towards the South Summit in May 1979. The West Summit, or Yalung Kang, is far more prominent at 8,505m, as there is a considerable drop to an intervening saddle.

---

6   Editor's note: many of the heights in this book are either disputed or uncertain. Uncertain heights are quoted first from Wikipedia and then from the *Alpine Journal* and *Himalayan Journal*. Heights which remain uncertain to more than 50m are preceded by 'c.'. Throughout the book, 'm' is used for vertical heights and 'metres' for horizontal distances.

7   There is considerable debate and uncertainty about the heights along the main Kabru ridge.

About two and a half kilometres west-north-west of Yalung Kang lies an outlier of the central massif, the impressive snow peak Kangbachen (7,903m), with its huge north-west face rearing high above the Ramthang glacier.

The mountains and glaciers on the west side of Kangchenjunga and the Singalila Ridge are practically the mirror image of those on the east. No one can pass by on the north side of Jannu (Kumbhakarna, 7,710m) without being staggered by the enormousness of its sheer granite north-west face, so steep that in the upper part there is hardly a trace of snow. Jannu from the south, however, presents itself as a mighty throne.

On the east side of Kangchenjunga lies Siniolchu (6,888m), described by Douglas Freshfield during his 1899 visit as:

> the most superb triumph of mountain architecture and the most beautiful snow mountain in the world, standing out from neighbouring peaks as does Giotto's Tower from the rest of the Italian domes and *campanili*.

Between Kangchenjunga and Siniolchu lies the twin-summited peak Simvu, with its West Summit at 6,818m and the East Summit only one metre lower. Thirteen kilometres south of Simvu stands the holy mountain of Pandim (6,691m); its north ridge descends towards the Goecha La (4,940m). Although prominent and relatively accessible, Simvu remains unclimbed, thanks to a recent ban: to ensure their wellbeing, local villagers do not want the mountain's deity to be disturbed.

One ridge leading north-west from Kangbachen includes Ramthang Peak (6,601m): it terminates dramatically at Wedge Peak (Chang Himal, 6,802m) above the Kangchenjunga glacier. Directly west of Kangbachen are several other snow-fluted summits, of which White Wave, now with the official name Anidesh Chuli (6,950m), is the highest. The most prominent mountain on this ridge stands above the village of Kambachen on the Ghunsa Khola: Mera Peak (6,334m).[8]

Frank Smythe pointed out in the *Himalayan Journal* that not only is Kangchenjunga a complicated massif, but:

> direct access to the summit via a continuous ridge, which does not involve traversing any of the remaining summits, is only possible via the NE spur and the NNE ridge.

---

8   Not of course to be confused with the popular trekking peak near Everest.

Kangchenjunga radiates subsidiary ridges. Four glaciers are contained between them; they are arranged like a cross on the cardinal points. Of these, the Zemu, Kangchenjunga and Yalung glaciers all originate from the flanks of Kangchenjunga itself and the Talung from Kangchenjunga South Peak.

Of all the glaciers, the Zemu, at twenty-six kilometres, is the longest in the Eastern Himalaya, although it has been shrinking fast in recent years. The Zemu and the Talung, to the south-east, both feed into tributaries of the Tista.

The eastern versant is spread across the upper basin of the Tista. The southern border of Sikkim is defined by the youthful mountain tributaries of the Tista, the Rangit and Rangpo. This country is nearly all mountainous, the only flat land lying to the north where the debris from Himalayan glaciers and former ice sheets has created a levelling effect.

The climate of the Himalaya

The formidable height and length of the Himalaya affect the climate substantially: these mountains are a major obstacle to the passage of air currents. During the winter, the Himalaya protect the Indian subcontinent from the cold, continental air mass that lies over the great plateau of Tibet. This dense air would naturally gravitate from high to low pressure, towards the region of warmer, rising air from the Ganges basin; the Himalaya largely prevent this happening.

Similarly, in summer, the warm, moist monsoon air mass which moves up from the south-west, bringing rain to the Ganges plain, is checked by the three parallel ranges of the Himalaya and, as the monsoon is forced up and along the flank of the Himalaya, cooling with an increase in precipitation takes place, falling as snow above about 1,500m. By the time this summer monsoon has reached the watershed, air currents continue on to the largely moisture-free plateau. The currents have been drawn north because of the summertime low-pressure system prevailing over the scorching upland pasture and semi-desert of Tibet, in the Himalayan rain-shadow.

The amount of snow falling on the Eastern Himalayan mountains during the onset of the monsoon is massive. The south-west monsoon-bearing airstream is channelled towards the north-east of the subcontinent and brings far greater annual precipitation to this area than to the north-west. Darjeeling, for example, receives about 305 millimetres of rain per annum, with most of it falling during the four months of the monsoon: June, July, August and September. Towards the north-west, the average yearly rainfall drops to about 150 millimetres at Simla (Shimla).

During the monsoon period, massive problems can arise in travelling between villages, particularly on newly excavated 'jeep' tracks which invariably become blocked by landslips for the duration. Loss of life in farms and villages swept down the hillsides is not uncommon, especially in East Nepal and Sikkim.

The vast length of the Himalaya, coupled with the greatest vertical interval of any mountain range, plus the complexities of its topography, ensure there is a wide variety of climatic type from sub-tropical in the southern foothills to cold desert conditions on the northern flanks and the Tibetan plateau.

From a mountaineer's perspective, the main problem in climbing high mountains is always the wind. In the Himalaya, the prevailing winds at high altitude are the westerlies; the higher the climb, the fiercer the wind as the climber heads into the jet stream. Fortunately, this westerly airflow is interrupted in midsummer by the approaching monsoon. Until then, the westerlies can be so violent above 6,500m as to destroy tents, or even blow flaky stones uphill.

The full monsoon's imminent arrival, usually in early June, is marked by a relative calm, often after the middle of May. It seems the monsoon air mass simply checks the relentless path of the westerlies until early September. This period is the best time to climb big mountains. In fact, Kangchenjunga was first climbed on 25 May 1955. Our British team made the third ascent on 16 May 1979, followed by several days of perfect weather.

The climate of the Kangchenjunga massif

The north–south alignment of the Tista Valley allows the monsoon rains to penetrate far towards the north. The monsoon is particularly heavy in the north-east of the Himalaya and in Sikkim: in ancient texts, this tiny kingdom was known as Rong-yul, 'the Land of Gorges'.

Of all the 8,000m mountains, Kangchenjunga is most affected by the monsoon. It is this heavy precipitation, falling as snow, that has made Kangchenjunga such a dangerous mountain to climb: more avalanches pour down its flanks than on any other of the great Himalayan peaks.

After the monsoon has run its course, the westerly winds return to hammer the mountains above about 7,500m. The range of temperature is just as impressive as the precipitation, especially in winter, when temperatures in the south may remain higher than 30 °C, while in the north they may drop to below -30 °C, with an added wind-chill factor, particularly on the west side of the mountains.

Kangchenjunga's situation, south of the main Himalayan divide and rising from deep, humid subtropical valleys, means there can be heavy falls of snow at any time of the year. Based on the last hundred years or so of mountaineering on Kangchenjunga, the most favourable weather period for climbing is generally the middle of May, especially on the west side. Before the monsoon arrives, there is an opportunity to climb in calm, clear weather, although for how long and exactly when adds great uncertainty for expeditions climbing high in this region. In 1979, we were convinced that the only time for an attempt came before the beginning of the summer rains, the so-called 'pre-monsoon' period.

Whichever period is chosen, the mountains will make their own weather. The climate will always be the main challenge; expeditions must plan for the uncertainty of the arrival and departure of the monsoon and never underestimate its severity. Those expeditions during the post-monsoon period must plan for an erratic end to the monsoon. There can be no set date, and neither can it be known what the effect of the monsoon will be on the approach march: torrential flooding may sweep away jeep roads and hiking trails, while on the mountain there may well be huge falls of avalanche-prone snow, especially on Kangchenjunga, long recognised as the most avalanche-active of mountains.

The monsoon air mass does have one beneficial effect in mitigating the severity of the westerly winds. This may not be so relevant on the east or lee side of the mountain, but it is a critical factor when climbing in the western quadrants during the pre-monsoon season. How much more of a challenge the Himalaya are to a mountaineer than the Alps, where there is one long summer season: in the Himalaya, as Frank Smythe reflected after his attempt from the north-west in the spring of 1930, there are only two openings:

> Both are pitifully short, and the total time available for attacking Kangchenjunga is no more than four or five weeks, and even this is liable to be interrupted by local bad weather.

Flora and fauna

> If there are no trees, there will be no water whenever one looks for it. The watering places will become dry. If forests are cut down, there will be avalanches. If there are many avalanches, there will be great accidents. Accidents also destroy the fields. Without forests, the householder's work cannot be accomplished. Therefore, he who cuts down the forest near a watering place will be fined five rupees.
> *Fourteenth Edict of King Ram Shah (1606–1636)*

The Himalaya must have been such a beautiful place 60,000 years ago. One could have walked through groves of *Cedrus deodara* (Himalayan cedar), some of them even today sixty metres high with a girth of three metres. (The name derives from the Sanskrit *devadaru*, meaning the 'wood of the gods'.) The magnificent trees found in remnants of ancient forest which have survived mankind's need to cut them down indicate how these forests might have looked.

Pockets of virgin woodland survive, protected from earliest times by nature-worshipping forest folk who still regard them as sacred. Today, a growing number of natural woodlands are protected as wildlife sanctuaries. The Kangchenjunga National Park in Sikkim is home to many high-altitude birds and animals, such as the pheasant and woodcock, the musk deer, the red panda and the clouded leopard. Their habitat is coniferous forest, alpine shrub and meadow, a pristine sanctuary for all sentient beings – mankind included – to find spiritual renewal.

Most of the original inhabitants of India and the Himalaya enjoyed a close affinity with the forest. This strong concern became apparent during the late Vedic period, when forests were being cleared to make way for agriculture: dependence on nature and the forests was recognised as being of fundamental importance. Good forest practice became established, with a person appointed to punish those who took more than their allotted share of timber. This system remains in operation today in some rural areas of Nepal and Sikkim.

Foreign and indigenous authors refer extensively to the vegetation of Nepal and Sikkim. Of the foreign publications, Sir Joseph Hooker's *Himalayan Journals* (1854), *Nepal* by Toni Hagen (1960), *Himalayas* (1988) by Augusto Gansser (and co-authors), and some very good, short but comprehensive accounts in such trekking guides as *Nepal* (1990) in the Lonely Planet series all stand out. A charming Nepali discussion on growing things in *Sacred and Useful Plants and Trees of Nepal* (1978) by Trilok Chandra Majupuria of Tribhuvan University provides not just scientific data but much about the spiritual dimension of forest cover that may serve as:

> anodyne to beings, mentally distressed and spiritually depressed in the sordid milieu of the material world.

In *Plants and People of Nepal* (2002) by Narayan P. Manandhar, the reader will find that about fifty-four per cent of Nepal has some sort of vegetation cover. Forested area constitutes thirty-seven per cent, shrub lands and depleted forests five per cent and grasslands twelve per cent (Shakya and Joshi). It is estimated that there are about 7,000 species of flowering plants in Nepal, about 300 of which are endemic. The author conducted his own research on foot in all seventy-five districts of Nepal and describes the use of 1,517 kinds of plants belonging to 858 genera and 195 families, about one fifth of Nepal's flora. There is also an index of more than 6,800 plant names, vernacular and scientific.

The fauna and flora of the Himalaya conform to the variation of rainfall and temperature and to the nature of the soils and lie of the land. The diminished atmospheric pressure, coupled with very low temperatures, have led to mammals adapting to these extreme conditions, such as the yak, snow leopard, Himalayan tahr, blue sheep, ibex and Tibetan gazelle. Their existence above the treeline is in itself only possible because of the adaptation of different plant species to the low temperatures, rainfall, ultraviolet radiation and very short growing season.

Oak, birch and many species of rhododendron populate the higher regions. The rhododendron proliferates because the leaf litter forms a toxic ground cover inhibiting the growth of other species. Most visitors to Nepal will be grateful for this, especially in March and April when the forests are alive with its wonderful flowers. The chir pine, blue pine, silver pine and hemlock are well adapted to the highest elevations, where the precipitation is low and the temperatures plummet. Struggling beyond the pines is ancient, stunted juniper, wind-blasted as it crawls over the rocky ground towards the snow line.

The highest-growing plant species is a moss found at 6,480m. Below that level, the herbivores especially relish the perennial grasses which struggle on the thin soil, dotted with gentian, blue poppy, wild strawberry and, quite often, wild rhubarb, which grows especially well in the glacier ablation valleys, in particular on the approach to Kangchenjunga.

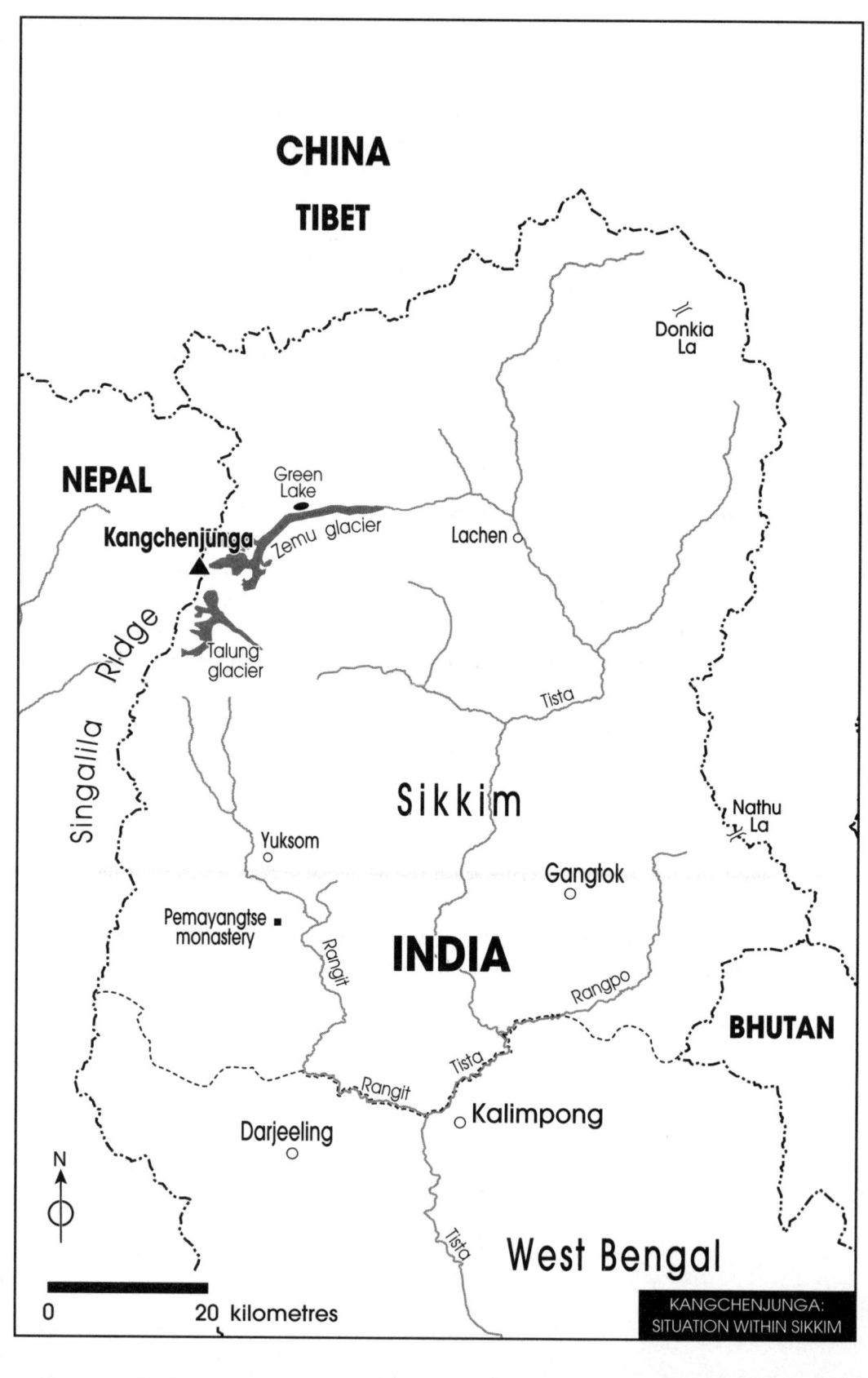

2

The Peoples of Eastern Nepal and Sikkim

Historical origins

We know very little about the original inhabitants of the Himalayan valleys. It is currently understood that there were Stone Age (*Homo erectus*) folk inhabiting North India, and the entire Himalaya, from about two million years ago.

It was during the Vedic or Iron Age period, from 1500 BC to 600 BC, that the foundations of Hindu culture were established, outward signs being the sanctification of the cow and the pipal tree.[1] During the late Vedic period, the centre of culture moved east into what is now Nepal and the Indian state of Bihar. The discovery of iron and its uses enabled this migration east through the dense forests of the Gangetic plains to the Nepali Terai (low-lying jungle country along the southern Nepali border area).

Thereafter, 600 BC to 300 BC encompasses the sway of the Mahajanapadas, the sixteen powerful kingdoms and republics which flourished across northern India to Bengal.

Ashoka (304 BC–232 BC)

This seminal Mauryan ruler established the hub of his empire at Palaliputra (modern Patna). At the greatest expansion of the Mauryan Empire by 265 BC, Ashoka's rule included all modern-day India as well as significant portions of Bangladesh, Bhutan, Nepal, Pakistan, Iran, Afghanistan and part of south-west China.

Ashoka took an active interest in preserving his nation's forests as well as fruit orchards, herbal gardens and vegetation which had medicinal properties. His concern for conservation extended to prohibiting hunting

---

1 Pipal (sometimes 'peepal') or bodhi tree: *ficus religiosa*. The 'sacred fig' has a religious significance in Hinduism, Buddhism and Jainism.

certain wildlife and forbidding cruelty to all animals. He included his ecological concerns in one of his edicts:

> Wherever medical herbs suitable for humans or animals are not available, I have had them imported and grown. I have planted mango groves, and I have had ponds dug up and shelters erected along the roads. I have had banyan trees planted on the roads to give shade to man and beast.

Ashoka's empire embraced only the southern region of what is now Nepal, but he did visit the Kathmandu Valley, then known as Nepala, which the Mauryan dynasty held in high regard on account of the quality of its woollen cloth. Here, Ashoka had four stupas constructed. To ensure that his influence, and that of Buddhism, extended into the Himalayan regions, he required his daughter, Charumati, to marry a local prince, Devapala; after his death, she had another stupa built in his memory near their home at Pasupati.

Gupta Empire (AD 320–AD 600)
This empire arose through the military successes of its first rulers, Chandragupta I, Samudragupta and Chandragupta II. They administered their empire from Palaliputra (Patna), as had Ashoka. They kept together this realm from the Indus to the northern boundary of the Himalaya by granting greater regional and local autonomy.

Delhi Sultanate (AD 1206–AD 1526)
Muhammad of Ghor was assassinated in 1206. He was replaced by his former slave, Qutb al-din Aibak, who was officially recognised as the first ruler of the Delhi Sultanate, a Muslim kingdom that would continue to dominate life in the north of the Indian subcontinent, including Nepal and Sikkim, for 300 years.

Eastern Nepal and its peoples
The Nepali scholar and statesman Rishikesh Shaha, in his 1982 paper on *Ancient and Medieval Nepal*, commented:

> with an area of 54,717 square miles and a population of 17 million [29 million in 2017], Nepal is hardly a tiny country. Seventy per cent of the member nations of the UN are smaller than Nepal in terms of population and over forty per cent are smaller in terms of area. Yet she is dwarfed by her two giant neighbours, India and China.

Nepal extends almost 800 kilometres from the heavily monsoon-affected Eastern Himalaya to the drier central Himalaya. What is of primary interest is the series of interconnected valleys and basins lying between the main Himalayan range and the middle Himalayan range: here, Nepal came into being and continues as the cultural centre of the country to this day.

The designation 'Nepal' was originally applied only to the fertile Kathmandu Valley and surroundings, an enclosed bowl of some 570 square kilometres, 1,200m up in the Central Himalaya. It may be that the name originated from the Sanskrit *nipalaya*, 'place at the foot of the mountains'. If correct, then the country's name illustrates how much the history of Nepal has been influenced by its topography.

'Nepala' may alternatively have been derived from 'Newala' with 'Nepal' as the historically final diminutive. (See *Forever Incomplete: The Story of Nepal*, 2013.) According to Mahendra Man Singh:

> the Kirantas ... named this valley 'Newala' which in the Kirat-Limbu language means 'a large plain surrounded by hills'.

Between the malaria-infested jungles of the Terai and the sparsely populated Himalaya, the land and climate are most favourable for human habitation. This remote and until modern times relatively inaccessible valley was nevertheless a haven for 'refugees' from north and south. Within the population we see a mixture of ethnicities from many Asiatic regions, Buddhist and Hindu, but also Animist and Bön.[2] These ethnicities have made Nepali culture unique and fabulously interesting.

Why would Nepal not be affected by the empires that have risen and fallen over the last ten millennia in the north-east of the Indian subcontinent? Nepal lies at the foot of the Himalayan mountains with relatively easy southern access. Despite malarial jungle along the northern edge of the Ganges basin, it remains less of an obstacle to access Nepal from India than from Tibet, although Tibetan influences have not been absent: the Tibetan term *Niyampal* can be interpreted as 'holy land'. Since Buddhism has been associated with Nepal since the fifth century BC, the name of the country may be derived from the Tibetan.

Scholars of the pre-history and history of Nepal describe the 'Nepal Mandala', an area consisting of mainly indigenous Newari people sharing and merging their cultural, religious and political aspirations with those of

---

2   Bön: the ancient, animistic, arguably shamanistic and pantheistic religion of Tibet and its proximate areas.

the many migrants into the Nepal Valley down the generations. The Mandala was often referred to, before Kathmandu became dominant, as the 'Nepal Valley'. Nowadays, it is generally recognised as lying between the Tibet border in the north, the kingdom of Kiranti to the east, the kingdom of Makwanpur to the south and the Trisuli river to the west. Beyond the Trisuli, the kingdom of Gorkha eventually grew strong enough to reduce the power, influence and autonomy of the Mandala collective through conquest in the eighteenth century.

The other feature of Nepal's history relevant to the Kangchenjunga area is that there is no hard, written historical evidence for those ethnic groups and dynasties which legend and folklore suggest existed before the Changu Narayan temple inscriptions of the Licchavi dynasty in the fifth century AD. Evidence of Emperor Ashoka's visit to the Kathmandu Valley around 265 BC, for example, during the reign of the fourteenth Kiranti king, Stunkho, is provided by his stupas around Pathan at the four cardinal points. (During the auspicious full moon in August, Nepali and Tibetan Buddhist pilgrims walk round all four stupas in a single day.)

The latest archaeological investigations indicate that areas of Nepal were inhabited by *Homo erectus* as far back as the Early Palaeolithic period from 200,000 BC to 50,000 BC. Stone Age tools have been discovered, many of them resembling those found in south-east Asia. This discovery suggests that at least one route taken during the original dispersal of human beings from East Africa to south-east Asia passed through mid-Nepal. It would have been a much safer option than attempting to penetrate the dangerous jungles bordering the Gangetic plain.

Only ethnic groups which settled in East Nepal and Sikkim since the Neolithic Period, which in Nepal is considered to cover from c.4000 BC to c.2000 BC, later than in Northern India, are looked at here.

Between Ashoka's visit and the fifth-century inscriptions, the evidence of political, economic and cultural life is still little understood except for what can be deduced from classical Indian and literary sources of Nepali origin. According to Rishikesh Shaha:

> A number of *vamsavalis* or chronicles give us some information about the origins and earliest history of Nepal. The oldest and most reliable of these is called the *Gopalarajavamsavali* [*The Chronicle of the Gopala Kings*].

(Shaha acknowledges that Dr Dilli Raman Regmi's researches have produced the most extensive source material for further research.)

Dor Bahadur Bista in *Fatalism and Development* (1991), has interesting comments on Nepal's pre-history as well as Nepal's struggle for modernisation:

> Till the first century AD, myths and legends are the primary sources of information on Nepal. The myths of the civilization of the Kathmandu Valley and the hills immediately around it provide evidence of the diverse origins of the people who converged and settled down within it. Other areas represented in mythic lore are Janakpur, the domain of King Janak, father of Sita in the *Ramayana*; Biratnagar in Moran, the country of King Birat of the *Mahabharata*; Kichakban and Kichak-day in Jhapa where King Kichak was killed in a duel with Bhimsen in the *Mahabharata*. All three areas are in the eastern Terai districts of Nepal.

As mentioned in the chronicle *Gopalarajavamsavali* (Vajracharya and Malla, 1985), the name of a group of Gopala cowherds who founded the first kingdom in the Kathmandu region is said to have been 'Nepa'. There is no agreement yet about the identity of the Gopala people beyond their being cowherds. Some historians suggest that the Gopalas had a connection with the Gopalas of the Yadava tribe from the Gangetic plains, associated originally with Mathura and Krishna, the Vais(h)navic deity also known as Gopala. This finds favour with Sanskrit scholars and Vais(h)nav pundits within Nepal. Others suggest that these early settlers were called Gopala simply because they were pastoral nomads, not because they were descended from the Gopalas of the plains (Panta, 1987). (*Gopala* is a Sanskrit word in which 'go' means cow, and 'pala' means 'herder' or 'keeper'.) The Gopalas who founded the first settlements in the Kathmandu Valley used a more primitive agricultural technology than that of the relatively advanced agriculturalists from the plains. The first attempts at agriculture in the valley seem more like the work of pastoral peoples such as the Kirat and the Khas.

Kiranti Empire (800 BC–AD 300)

The scholarly consensus is that the Newari were the original inhabitants of this 'Nepal Valley', with other ethnic groups who became the Kiranti (sometimes Kirat or Kiratas) people, a collective label which also refers to the Rai, Limbu, Sunwar, Tamang and others. The 'Kiratas' are referred to in the *Mahabharata* and in the *Atharvaveda* in the sixteenth century BC as skilled archers and Himalayan hunters or trappers. It seems the majority of the Kiranti people migrated to their present location via Tibet, Assam and mountainous Yunnan in south-west China and Burma some 10,000 years ago. It may have been much earlier, however, if recent interpretations of archaeological finds during excavations by a Russian–Nepali study programme are correct.

For almost three millennia, from at least the ninth century BC, most of the Eastern Himalaya has been the homeland of the Kiranti people. From 800 BC until AD 300 their domain included the Kathmandu Valley, where a succession of twenty-nine kings held sway. During the Kiranti rule, Buddhism took root in the country. It is possible that the Buddha himself visited the valley, staying in Pathan with his disciple, Ananda. Most of this long period of Nepal's history is otherwise not well documented.

The Licchavi

Many historical examples show how much the cultural and political life of Nepal has been influenced by India. Around AD 300, the Licchavi, the first significant dynasty in Nepal, emigrated from modern Bihar and conquered the country of the last Kiranti king, Gasti. They established their capital city at Deopatan, which became the important contemporary Pashupatinath temple complex. The Licchavi were renowned stonemasons. They produced fine sculptures, still seen in the temples of Kathmandu, particularly at Budhanilkantha.

Such kingdoms maintained quite clearly defined boundaries since they lay within the physical constraints of the surrounding valleys, hills and mountains. The Himalaya, while they may deny easy expansion, have sometimes helped to protect kingdoms from would-be aggressors.

The first physical record of the Licchavi is an inscription by Manadeva I in AD 464. It refers to three previous rulers, suggesting the dynasty began in the fourth century AD. Manadeva I was a strong leader: an inscription in the fourth century Changu Narayan temple, eleven kilometres east of Kathmandu, refers to his greatness. Throughout the valley, Manadeva II also left stone inscriptions extolling the virtues of his mother.

The next 600 years – and in particular the reign of the last Licchavi king, Jayakarma Dev – are considered a golden age in Nepal's history. This period became synonymous with strong decentralised government, peace, prosperity, justice, religious tolerance and the flowering of the arts.

Amshuverma ruled as *mahasamanta* (prime minister) from AD 595 to AD 621; he reduced King Sivadev I to a figurehead. He opened trading routes through the Himalaya to Tibet; the Buddhist religion followed. With the Newari clan at the centre of commercial and cultural life, trade flourished in metal, wood and stone handicrafts, as well as in paper from the *lokti* shrub which grew profusely throughout central Nepal.

Early in the first millennium, Newari merchants gravitated towards the important centres which grew up conveniently between the trading partners, India and Tibet. From Pathan (Lalitpur), commodities were sent north into Tibet. Buddhist lamas and Buddhism followed. The Newari, through their natural aptitude for business, gained considerable wealth for their nation at this time, from customs duties used to support the artistic heritage of the Nepal Valley.

During this period, the valley was transformed from a remote backwater into a major commercial and intellectual centre, with Buddhists and Hindus living in harmony. The pagoda evolved; it greatly impressed Chinese pilgrims. The most notable feature of such buildings was the most intricate woodwork carving imaginable. Such buildings are not only of religious significance but have become towering monuments to man's skill and his ability to achieve perfection by fashioning such harmonious proportions that they lift the spirits of all who behold them.

Several exchanges between the Licchavi and the Chinese occurred in the mid-seventh century. These later became irregular, owing to intermittent conflict between the Tibetans and the Chinese.

The Licchavi dynasty eventually disintegrated, weakened by attacks from the Pali dynasty of Bengal; at about that time, the Thakuri dynasty grew in strength and ruled Nepal from the ninth to the twelfth century until they, in turn, were superseded by the Malla dynasty.

The Malla dynasty (1201–1779)

'Malla' derives from the Sanskrit and means literally 'wrestler'. Mahendra Man Singh in *Forever Incomplete: The Story of Nepal* (2013) pinpoints when this moment of origin occurred:

> Raja Ari Deva, in about AD 1200, was engaged in this sport. Word was brought to him that he had been gifted with a son. Thereupon he proclaimed that his descendants would suffix the appellation 'Malla' to their names.

Thus began the longest-ruling dynasty in Nepal: it continued for almost 600 years. It was not an easy period, mainly because of Muslim invaders of India provoking the increased militarisation of adjacent kingdoms. The Khas kings from India, for example, seeking sanctuary, moved into western Nepal to avoid aggressive neighbours who also occasionally raided the Kathmandu Valley between 1275 and 1335.

In 1255, a third of the population of the valley, some 30,000 people, including King Abhaya Malla, perished in an earthquake that was unusually devastating because the epicentre lay just below the centre of Kathmandu. In the mid-fourteenth century, further Muslim invasions caused damage to Hindu and Buddhist shrines. During the final centuries of Malla rule, the Mughals grew in power and influence in India. They did not attack Nepal directly, but they did influence institutional life as many dispossessed rulers from northern India again sought sanctuary in Nepal. They brought advanced military technology, particularly firearms, and the Mughal dress code. Both are evident in contemporaneous portraits which show the Malla rulers posing proudly with their new weaponry.

The Malla kings succeeded, over many years, in keeping together the country centred on the Kathmandu Valley. In the fourteenth century, they organised the people into strict Hindu castes according to profession and ethnicity. Jaksha Malla extended his reign far beyond the Kathmandu Valley during his fifty years in power; in 1488, he divided the kingdom among his three sons, a most significant act with far-reaching consequences since, in division, the country became weaker and eventually was unable to withstand the Gurkha invasion of 1768.

The Malla period in the Kathmandu Valley was renowned for the originality and superb craftmanship of its buildings. Today, the centre of Kathmandu has the most remarkable monuments: shrines, pagodas and palaces. The majority were built during the reign of the Malla kings, particularly during the latter period of the three kingdoms: Kathmandu, Bhaktapur and Pathan. Although the three princes were often at war with each other they remained loyal to their main purpose, which was as protectors of the *dharma*.[3] They built numerous shrines and temples, all in competition with each other. By the mid-eighteenth century, King Jaya Prakash Malla had built Kathmandu's Kumari temple. This was followed by the five-tiered Nyatapola temple in Bhaktapur, literally the high point of pagoda-style building in the Nepal Valley.

---

3 *Dharma*: the Buddhist understanding of the laws of cosmic order.

By the sixteenth century, the Nepal Valley had become ever more important through offering two separate trading routes into Tibet, one via Banepa to the east of Kathmandu and the other through Rasuwa and the Kyirong Valley to the north. In early spring, Kathmandu and Pathan, as well as many of the surrounding towns, would be teeming with merchants who had arrived from India in the winter to avoid the malaria rife throughout the summer months in the Terai. By early summer, the passes into Tibet – and by implication those passing near Kangchenjunga – would be open for trade in musk, wool, salt, silk and other commodities.

The fertile soils of the Kathmandu Valley, in many parts producing two crops a year, were more than able to support three kingdoms. Newari legend has it, with geological evidence in support, that the Kathmandu Valley was once a huge lake. A team of geologists from Kyoto University has confirmed that 10,000 years ago, the lake drained away during a major crustal movement in the region. Hindus have attributed the formation of the valley to Vishnu, with one blow of his sword, cutting through the outer rim to allow the lake to empty. Another Hindu legend has Krishna (the physical embodiment of Vishnu) draining the lake by directing a thunderbolt at the containing hills to produce the Chobar Gorge, nine kilometres south-west of Kathmandu, through which the Bagmati river still flows.

The Newari

The Newari have a rich and varied history, receding into mythical times, reflected in numerous festivals and celebrations. Their legend that the Kathmandu Valley was formerly a huge lake goes some way to suggesting that their claim to be the original inhabitants after the lake emptied may be true. They give a Buddhist interpretation to the draining of the lake, in believing that the water was released by Manjushri, the Bodhisattva of Wisdom, who, with a single swipe of his sword cut through the surrounding hills to drain the lake and reveal a land of great fertility.

Ethnologists believe that the Newari are mainly of Palaeo–North Asian ethnicity with strong Tibetan connections. The Newari have embraced Buddhist, Brahminic and Hindu beliefs and customs, all related but by no means identical, a remarkable example of an ethnic group combining the divergent views of two great religions. They are widely known also for evolving a unique style of building which uses a combination of brickwork and timber frame. The pagoda style, which proved so attractive to neighbouring countries, was their hallmark. The positive result of the catastrophic earthquake of 1255 was to stimulate a boom in Newari-designed construction in the Kathmandu Valley. Nearly 800 years later, after the most recent earthquake in Nepal, there is a comparable increase, but very little exists of today's construction that anyone would want to preserve!

Araniko (1245–1306)

The leading Newari architect in Nepal was Araniko, born in the Kathmandu Valley during the reign of Abhaya Malla. He died in China in 1306, aged sixty-two. He was given high honours in China. The stupa in the Miaoyang monastery in Beijing is one of his most famous achievements. Araniko's legacy is enshrined in his ability to integrate the artistic traditions of Nepal with Chinese Buddhist art. The Nepalis, in honour of their thirteenth-century architect, named the road linking Kathmandu with Kodari on the Tibetan border the 'Araniko Highway'. (It was built in the 1960s and has proved to be one of the most dangerous roads in the world. It follows an old yak and pack-animal route along the Sunkosi river, through the Himalaya, to link with the Kodari–Lhasa road passing quite close to the Sakya monastery.)

The Newari are not easily able to trace their ancestry to a common source. In recognising the diverse customs that have coalesced to become the Newari of today, ethnologists have labelled them a community with 'relic identity' of religion, ethnicity, language, caste and other social structures. They are inextricably associated with the Nepal Valley, the enclosed basin at the source of the Bagmati river, so much so that Harka Gurung, one of Nepal's foremost ethnologists, believes that 'Newari' is a cognate of 'Nepal'. Most historians believe that the Newari have lived in the Nepal Valley since at least the seventh century BC.

Being essentially traders and artisans, the Newari are a very sociable people, and to this day live in densely packed communities of tall, multi-storeyed tenement-like buildings above narrow alleyways. They are confirmed urbanites, quite unlike the majority of Nepal's hill people who have steadily outnumbered them since the Maoist insurgency.

What strikes many visitors to Nepal is the vibrancy of communal life. No town resonates more with life than Kathmandu and, in particular, Thamel, so full of commercial energy, thronged with largely good-humoured locals and visitors, buying and selling goods and services much as it was at the beginning of Newari enterprise. Among this bustling throng is found everyone from *sadhus*, religious ascetics who have completely renounced the worldly life, to rickshaw wallahs straining the pedal and chain of their two-wheeled conveyances to transport local people, tourists and merchandise between factory and shop. So much industry has built up that it has spilled across the former paddy fields into the surrounding villages as cottage industries. Handicrafts in wood, metal and stone are still being produced, just as they have always been manufactured in the valley by skilled artisans of the Newari clan.

The contemporary urban dynamism of the Newari is evident not only in the capital but also at busy commercial manufacturing towns along trading routes in Nepal, such as Pokhara. The Newari are also to be found in Bihar, Bengal and Sikkim, at the centre of trade and industry. Just before the Chinese overran Lhasa in 1950, around a thousand Newari businessmen and their families had to quickly pack their bags and move south to Nepal.

The Newari are now, however, a minority ethnic group, since so many hill people have flooded into the Kathmandu Valley following the Maoist insurgency of 2005–2010.

The Newari have underwritten the visible culture of Nepal and the Eastern Himalaya through their commercial knowledge and artisan skills. They usually congregate wherever there is an opportunity to trade, but also where their craftmanship might be appreciated. They may have left Lhasa and other towns in Tibet in a hurry, but Newari communities still thrive in Sikkim and northern India.

The Limbu and the Rai

These peoples form the major ethnic group of the present Kiranti community. They are the earliest known inhabitants of the Himalaya east of the Arun river. Today, most of the 700,000 Limbu still live between the Arun and Mechi rivers, although large communities of Limbu live nearer Kathmandu as well as in northern Bengal, especially around Darjeeling. Close to the Kangchenjunga massif and the Singalila Ridge reside some quarter of a million Limbu, mostly in the area known as Limbuwan.

The most important summit for the Limbu seems not to be the highest peaks of Kangchenjunga but its steeply dramatic satellite peak, Jannu. (To the Limbu, Jannu is called Phoktanglungma, sometimes Phoktakunga or Photaklungma.) Here, according to their ancient religious texts, the gods meet, as they have done since creation. On the summit of Phoktanglungma the three creator and developer gods gathered for the purpose of creating human beings.

The history of the little-known Limbu people is an inspiration to all who value freedom of expression. Despite being indigenous to East Nepal and West Sikkim, the Limbu have suffered suppression of their language, culture and beliefs, centred on worship of the goddess Yuma. They have been oppressed repeatedly, since the seventeenth century in Nepal, at the hands of invaders from Gorkha, and in Sikkim from Tibetan lamas, especially at Pemayangste monastery, as well as from state administrators in Sikkim. They thus suffered oppression by both secular and monastic forces.

The peoples of Sikkim

The topography of Sikkim has ensured that until recent times it enjoyed a unique history and culture largely independent of Nepal, Tibet and Bhutan.

Its mountains and jungle have to a large extent insulated Sikkim from religious and cultural influences from the south. The destiny of Sikkim had been guided by Guru Rinpoche's travels throughout the region during the eighth century AD, but the early history of Sikkim before the seventeenth century is otherwise obscure. However, in the thirteenth century, a legend states that Guru Tashi, a Tibetan prince, had a vision that he and his family, including his five sons, should travel to the south. Eventually, the family settled in the Chumbi Valley, a Tibetan salient that protrudes south between Sikkim and Bhutan. After Guru Tashi's death, his eldest son, Khye Bumsa, established contact with the Lepcha patriarch and holy man Tetong Tek, on the other side of the Donkia Range at Gangtok. They became firm friends, resulting in an agreement that brought about ties of brotherhood between the Lepcha and Bhutia people, who were beginning to infiltrate Lepcha regions. The Lepcha writer A.R. Foning records:

> Eventually, as a result of their friendship pact, we agreed innocently to accept Kings among us ... They were installed in 1642 at Yoksam with our consent.

The Lepcha had never previously looked to any authority figure other than the elders of their tribe.

A large number of Nepalis migrated into Sikkim during the eighteenth and early nineteenth centuries after Prithvi Narayan Shah unified Nepal by armed conquest, following the Anglo–Nepali War of 1814–1816. The lives of the indigenous Lepcha people, as well as other ethnic groups such as the Limbu, Rai and Bhutia, were drastically altered. Statisticians in India and Sikkim do not seem to have produced a demographic breakdown of the ethnic groups comprising the Nepali emigrants to Sikkim, but without doubt those Nepalis engaged in trade, crafts or building are likely to have been Newari.

The Lepcha

The other aboriginal people who occupy the high valleys of north-west and central Sikkim, near Kangchenjunga, are the Lepcha. Their strong oral tradition, however, does not refer to migration. It is probable, therefore, that they are indigenous to this area.

Their language belongs to the Tibeto–Burman family. Their genetic characteristics suggest they are of North Asiatic origin, although the Lepcha themselves firmly believe they are living in their original homeland. They have certainly lived in sight of the Kangchenjunga massif for a very long time.

The Lepcha fared no better than the Limbu, despite having occupied land for centuries before the arrival of the Bhutia (Tibetans) in the seventeenth century. This very turbulent period was ended by the acceptance of Phuntsok Namgyal as King of Sikkim in 1641. At his coronation, an agreement, known as the Lho-Men-Tsong-Sum, established an alliance between the Bhutia, the Lepcha and the Limbu. All three clans were given equal participation in the governance of the country; that the first three *chogyals* took Limbu and Lepcha wives demonstrates, at least initially, the good intentions underpinning this agreement. By 1741, however, the lamas from the civil administration had banned all support for the Limbu language. Despite secular and monastic opposition, the Limbu kept their language and culture alive in their homes.

By 1914 a linguistic revival had begun, but it was not until 1985 that the Limbu language was officially recognised as one of Sikkim's state languages. In 2007, the Limbu were finally recognised as a listed tribe by the government of India, an event of great benefit to the preservation of their cultural heritage. The fascinating if chequered history of the Limbu is well told in Pema Wangchuk and Mita Zulca's *Khangchendzonga: Sacred Summit* (2007). It assesses the significance of the Limbu saint Teyongsi Srijanga and explains why a twenty-six-metre statue of him was unveiled in 2018, near to where he was martyred, in sight of Kangchenjunga. The birthdate of Teyongsi Srijanga is now observed as a state holiday.

A Limbu study centre and museum at Hee Bermiok Hill is planned. This will help the dissemination of information about the Limbu people and give them recognition and encouragement to preserve their unique culture.

The Limbu have, according to J.R. Subba in *Mythology of the People of Sikkim* (2009), given the name Senje Lungma or Sesey Lungma to Kangchenjunga, where they believe their omnipotent goddess, Yuma Sammang, resides.

The whole of Sikkim has long been considered by the Tibetan Buddhists to be a sacred country, designated a *'beyul'* or hidden land, a place of sanctuary in troubled times.[4] Sikkim did escape the turmoil that afflicted vast areas of northern India under the all-conquering Muslims. Protection was afforded by the Singalila chain of mountains and by the mountain barrier, Kangchenjunga, as invaders could penetrate no further east. The western side of Sikkim was always considered to be at the heart of this *beyul*. It is, after all, where life-giving water originates from the soil to support the sequestered Lepcha people. Even today, they know it as *Nye Mayel Lyang* ('Land as Pure as Heaven').

The Lepcha are thought to make up fifteen per cent of Sikkim's population. Their language and script are derived from Tibetan. Most of the Lepcha have adopted the Tibetan Buddhist faith from immigrant Bhutia people. A sizeable number have converted to Christianity. Underlying all their beliefs is this people's profound knowledge of the land and its ecology, particularly the properties of local plants. A strong indication of their retaining an animist past comes from their offerings to nature spirits.

From the few Lepcha I have met, they seem to resemble the Nishi, who helped Greg Child and me survive three weeks in their forest in Arunachal Pradesh. They too are a kind and gentle people, completely attuned to life in that dense, wet woodland. The Lepcha are wonderfully dextrous and are said to be magicians with bamboo crafts, as are the Nishi, who can weave lengths of sheared bamboo so tightly their baskets can hold water, while their home-made sou'wester-shaped headgear keeps out the rain. The handles of their machetes are of woven bamboo, while their arrowheads are secured with bamboo woven so tightly that it never moves; all most aesthetically pleasing.

Some Lepcha continue to practise a shamanistic religion known as Mun. They often merge ancient and modern beliefs and join the Bhutia in their rites and ceremonies to pay respect to the deities who reside on the snowy summits, particularly on Kangchenjunga.

The Lepcha generally lived their lives in harmony with each other, a necessity for surviving their hunter-gatherer, slash-and-burn existence. Later they developed into subsistence farmers, with the Lepcha and the Limbu combining in a federation to avoid inter-tribal warfare. The happy arrangement of all three ethnic groups only lasted a few decades, however, when the Limbu and Lepcha retreated to their homes in the high mountain

---

4   A curious modern 'take' on the *beyul* as a portal to a spiritual sanctuary can be found in Thomas Shor and Tenzin Palmo's *A Step Away from Paradise* (Penguin India, New Delhi, 2011).

valleys radiating from the Kangchenjunga massif. There, these peace-loving people, at one with the land, could continue to live in unusual harmony with each other.

Kangchenjunga, so dominant in the lives of the Lepcha, will always be a reminder of a slower, more meaningful way of life. They have their own name for Kangchenjunga: *Kingstsoom Zaongboo Chu*, 'The bright auspicious forehead peak' that welcomes the new day, as they wait, deep in shadow on the valley floor, for the early morning sun to turn the cold grey snowy summits golden. Kangchenjunga reigns supreme for the Lepcha. It is the place where the first Lepcha couple, Tukbothing and Nazong Nyu, were created by Itbu Mu from the virgin snows of Kangchenjunga's summit – it is their own Garden of Eden story.

PART 2
Western Approaches

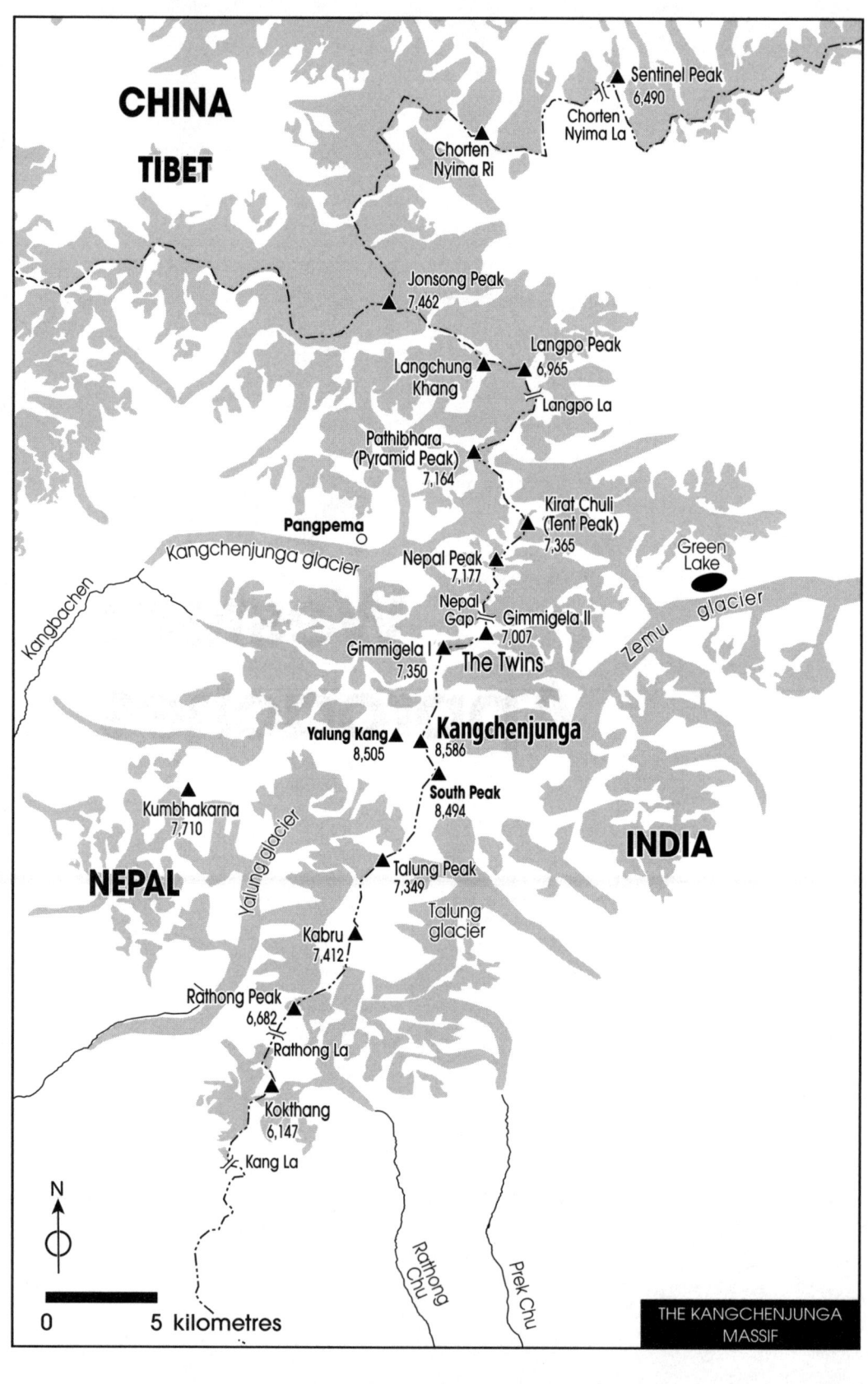

3
Missionaries, Traders and Politicians

To locate the legendary patriarch Prester John, thought to rule over a lost nation of Nestorian Christians cut off by evangelic Muslim hordes, was one of the two main motives for Jesuit and other missionary priests to travel across the Himalaya and high mountain deserts, without map or much information. They faced dreadful hardship: freezing weather, periods of drought, stupendous gorges and high passes where the oxygen was one third less than at sea level; local rulers who might forbid progress, or bandits who might rob or kill them. There is some evidence for the latter, but more missionaries died from illness or sheer exhaustion.

The main motive for their remarkable journeys was to convert the heathen to Christianity. Strangely, they were least successful in the countries which made them most welcome and gave them free rein to practise their calling. The tolerance displayed by the people of the Himalaya and Tibet towards the Jesuits largely derived from their confidence in the efficacy of their own faiths. The natural curiosity of the host community to know about the Christian faith also contributed. The Jesuits enjoyed only limited success in Asia compared to the converts and influence they gained in Europe and the New World.

Very little was known about the Himalaya and Tibet before Father Jean-Baptiste Du Halde's *Description géographique, historique, chronologique, politique et physique de l'Empire de la Chine et de la Tartarie Chinoise* (1735) and Jean-Baptiste d'Anville's *Nouvel Atlas de la Chine, de la Tartarie Chinoise et du Thibet* in forty-two sheets (1737). The Jesuit and Capuchin missionaries' written work, however, was seldom published outside ecclesiastical circles.

Several missionaries reached Central Asia after the Franciscan friar Odoric of Pordenone (1286–1331) first travelled through the area. Two such travellers headed north from Bengal to become the first Europeans to reach

Bhutan, from the western part of which Kangchenjunga is clearly visible. No one knew of these journeys except their superiors in Rome and a small number of correspondents.

Father Estêvão Cacella (1585–1630) and Father João Cabral (1599–1669)

The Portuguese Cacella had spent time at the Tsaparang mission in western Tibet before setting off with his younger colleague Cabral to establish a second mission station at what is now Shigatse.

They left Hooghly (Hoogli), the Portuguese mercantile town on the Ganges delta, in August 1626. They travelled through the then independent kingdom of Cooch Behar to reach Bhutan in February 1627, continuing via the Raidak Valley to reach Paro Dzong on 25 March. They experienced tolerance and hospitality during their nine months in the country: the Dharma Rajah offered to provide permanent accommodation for the fathers and even to have a church built for them.

Cacella, however, decided to leave Paro to fulfil the main objective of their journey. He travelled into Tibet via the Chumbi Valley and after twenty days reached Shigatse, where he was well received by the king of Ü-Tsang.[1] Cabral eventually joined Cacella on 20 January 1628. There is no record of their exact route but the obvious way from Paro would be via Phari in the Chumbi Valley, then through Gyantse to Shigatse and the famous Tashi Lhunpo monastery.

The king gave them permission to preach and made a suitable residence available, with ample provisions and a servant to help them become familiar with Tibetan customs. This favourable situation did not last; eventually, the priests left Shigatse, with Cabral leaving first, for India via Nepal.

This intrepid pair had become the first Europeans to visit Bhutan and to cross the Himalaya in winter conditions. Cabral became the first to traverse Nepal. He was helped on his journey with letters of support from the king of Ü-Tsang, which prompted the king of Nepal to extend this royal help all the way to Patna, on the Ganges.

The mission collapsed after only a few months, mainly due to lack of funding and Cacella's poor health. Both men had sent reports to their provincial superior, Father Alberto Laertius, senior Jesuit at Malabar on the south-west coast of India. These reports contained interesting observations on the religious customs, cultural and political life of Bhutan and southern

---

1 Ü-Tsang: one of the three historic provinces of Tibet.

Tibet but, apart from a brief observation that it was very cold, and that the rivers froze in January, little else was recorded about the 'Snow Mountains' along their route. In fact, the outside world knew nothing of these travels as their reports remained locked in the Jesuit archives in Rome. The Dutch Jesuit historian Father Cornelius Wessels discovered their letters in the 1920s. They provided only minimal detail but contained useful dates. The reports do refer to Nepal by name, the first reference to it by Europeans.

Ippolito Desideri (1684–1733)

The first Western traveller to reach close to the Kangchenjunga area was a twenty-seven-year-old Italian priest, Ippolito Desideri. He volunteered to open a mission in Tibet seventy years after the mission at Tsaparang collapsed. Having received the blessing of Pope Clement XI, Desideri set off from Genoa on 22 November 1712 on an epic journey across the oceans and northern India, then across the Himalaya and so to Lhasa, where he spent five years before returning via Kathmandu and Nepal, to arrive on 23 January 1728 in Rome, which he had left fifteen years and two months earlier.

The importance of Desideri's journey cannot be overestimated. His wonderful travelogue was kept from the world until 1875, when his manuscript was discovered among the papers of Cavaliere Rossi-Cassigoli of Pistoia, Desideri's home town. The manuscript was not published until 1904, however: as a result, commentators on the Jesuits in Asia, and of Desideri in particular, have much distorted their narrative or failed to acknowledge Desideri at all.

Despite such a long delay in the publication of Desideri's 'An Account of Tibet', a considerable amount of information had become available to diligent researchers. For example, the final volume of Du Halde's *Description de l'Empire de la Chine* (1735) includes reports of the Jesuit journeys, as well as d'Anville's *Nouvel Atlas de la Chine, de la Tartarie Chinoise et du Thibet*.

The groundless accusations and insinuations by superficial authors on Tibet such as Markham, Waddell, Sandberg, Landon, Rawling, Ryder, Bailey and even Sir Charles Bell, should be known by everyone now researching Desideri's life. Fortunately, Father Wessels sets the record straight in *Early Jesuit Travellers in Central Asia, 1603–1721* (1924). He concludes with the observation that Sven Hedin, in his *Southern Tibet*, appreciates Desideri as a major contributor to our knowledge of Tibet:

> Add to this the general merit of his narrative, the absence of fantastical speculation, the quiet matter-of-fact way in which he

gives his observations, and nobody will call it an exaggeration if I regard Ippolito Desideri as one of the most brilliant travellers who ever visited Tibet, and amongst the old ones, by far the most prominent and the most intelligent of all. No candid reader of the preceding pages will demur at this tribute.

In researching this period of Tibetan history, the student will not be disappointed to read *An Account of Tibet: The Travels of Ippolito Desideri of Pistoia, S.J., 1712–1727*, edited by Filippo de Filippi with an introduction by C. Wessels (1937, revised edition). The reader can follow Desideri's 'solid piece of work' on how he and his companion, Freyre, made the first European crossing of the Zoji La (3,529m) en route to Leh, then followed the Tsangpo to Shigatse and so to Lhasa. He and Freyre were consequently the first Westerners to see that holiest of mountains, Kailash, and witness the pilgrims prostrating themselves along the *kora* route for days at a time.[2]

Freyre bade farewell to Desideri: having found no Capuchin monks in residence, within a few days of arrival he left Lhasa for India by way of Nepal. His health had been weakened by many years in the tropics and was now impaired further by the bitter cold and thin air of the Tibetan highlands.

Desideri at that time lived as the sole European in the Tibetan capital, where he was accepted as a priest. After five years, he was obliged to leave as this mission field had been allocated to the Capuchins. He enjoyed a better understanding of all things Tibetan than any other European at that time.

Desideri did impart his knowledge of Tibet to the Capuchins, helping them to a better understanding of the country and the people before he reluctantly left Lhasa for Nepal, via Gyantse, accompanied by the Capuchin friar Joseph Felix da Morro, on 25 April 1721. They crossed the Himalaya by the well-used 'Kuti road', which runs through Nyalam into Nepal. Desideri remained in Kathmandu for only a brief period, although long enough to have formed an opinion on the indigenous Newar people:

> These Neuars [Newars] are active, intelligent, and very industrious, clever at engraving and melting metal, but unstable, turbulent and traitorous. They are of medium height, dark skinned and generally

---

2   The *kora* is the circumambulation of Kailas(h) (6,638m), for religious purposes. The outer route is an easy if high trek of about fifty-two kilometres, approaching about 5,800m at the Drölma La. There is a shorter but harder inner route, which carries consequently greater religious merit. The circumambulation is normally followed clockwise, though Bön adherents perform the *kora* anti-clockwise.

well made, but nearly all bear deceit written on their faces ... They are cowardly, mean and avaricious ... are dirty in their habits. They are very superstitious in all things, futile observers and utter heathens.

Jean-Baptiste Du Halde (1674–1743)
Du Halde was a French Jesuit historian specialising in China. He collected information from seventeen Jesuit missionary reports which enabled him in 1735 to present the most comprehensive account of the Chinese Empire, the *Description de l'Empire de la Chine*, in four volumes, published in France. The following year it was published in the Netherlands, and in English in 1738 as *The General History of China*. Many more translations followed.

Voltaire said of Du Halde's work:

> Although it is developed out of Paris, and he has not known the Chinese, [he] gave on the basis of the memoirs of his colleagues, the widest and the best description the empire of China has had worldwide.

For the first time on a map, Chomolhari (7,326m), on the Sikkim–Bhutan border, was marked, as well as the high mountains of Sikkim, to which the name *Rimo-la* was given. As far as is known, this is the first reference to Kangchenjunga from beyond the immediate environs of the mountain.

Jean-Baptiste d'Anville (1697–1782)
This renowned Parisian cartographer, with a passion for accuracy, took nothing for granted: blanks replaced what could not be verified. He revised the maps produced by the Chinese and Jesuits, resulting in the well-known series of 1733, later re-published as the *New Atlas of China* (1737). D'Anville's relatively reliable map alerted the wider world for the first time to the vast extent of the Himalaya.

British trading missions beyond India in the seventeenth, eighteenth and nineteenth centuries
In Richard Hakluyt's *Voyages* in 1599, Europeans received their first published information about those who came from the 'other side' of the mountains:

> There is a country four days' journey from Cooch Bihar which is called Bootanter [Bhotanta] and the city, Hoottea, the king is

called Durmain [Dharma Raja], the people whereof are very tall and strong; and there are merchants – who come out of China and out of Muscovia or Tartary; they come to buy musk, cambals, agates, silks, pepper and saffron of Persia.³ The country is very great. Three months' journey. There are very high mountains in this country and one of them so steep that when six days' journey off it, one may see it perfectly ... those which come from the other side of the mountain, which is from the north, say there it is very cold, the northern merchants are apparelled with woollen clothes and hats, white hosen clothes and boots which be of Muscovia or Tartary.

Ralph Fitch (1550-1611)

Fitch is known to have reached Kuch (Cooch) Behar and therefore must have seen the Himalaya. In doing so, he would have been the first Englishman to set eyes on Kangchenjunga. He was certainly the first Englishman, if not European, to approach the 'Great Snowy Hill' by which other European travellers later denoted Kangchenjunga when viewing it from the plains of Bengal.

Warren Hastings (1732-1818)

Hastings's principal mission was to promote trade between India and its surrounding kingdoms and, if possible, with China through Tibet. Such trade was frequently interrupted and finally ceased with Prithvi Narayan Shah's successful four-year attack on the three Malla kings of the Nepal Valley in 1769.

In 1768, the Bhutanese appointed the aggressive Deb Judhur to be Deb Raja of the country. Deb Judhur overran Sikkim and then Cooch (now Kuch) Behar in 1772. This independent princely state appealed to the British for help. In 1773, the rajah and British troops forced the Bhutanese out of the Duars, the approach valleys, and back into the foothills of Bhutan. The Zhidar (ruler) of Bhutan, Druk Desi, returned to face a civil war in which he was defeated and forced into exile in Tibet, only to be imprisoned by the sixth Panchen Lama. The Lama then sent a letter to Hastings proposing a cessation of hostilities between Britain and Bhutan in return for mutual friendship. Hastings seized the opportunity, writing to the Panchen Lama to suggest 'a general treaty of amity and commerce between Tibet and Bengal'. He quickly followed up on this opening of relations by putting together

---

3   Cambal (sometimes 'kambal'): a shawl made of coarse wool.

a fact-finding mission to 'chart the unknown territory beyond the northern borders of Bengal'. The British could now gain first-hand information about the true configuration of the territories north of Company lands and of the people living among the snowy mountains visible from the plains of Bengal. Highest among them, to be identified later as the third-highest peak in the world, was Kangchenjunga.

George Bogle (1746-1781)

Hastings selected Bogle, a Scot from Daldowie in Lanarkshire who, at twenty-eight, had established a reputation for high intelligence and 'coolness and moderation of temper which he seems to possess in an eminent degree'. Hastings gave his agent the following instructions, dated 13 May 1774:

> I desire you will proceed to Lhasa ... The design of your mission is to pen a mutual and equal communication of trade between the inhabitants of Bhutan [Tibet] and Bengal, and you will be guided by your own judgement in using such means of negotiation as may be most likely to effect this purpose. You will take with you samples, for a trial of such articles of commerce as may be sent from this country ... And you will diligently inform yourself of the manufactures, productions, goods, introduced by the intercourse with other countries, which are to be procured in Bhutan ... The following will be also proper objects of your inquiry: the nature of the roads between the borders of Bengal and Lhasa, and of the country lying between; the communications between Lhasa and the neighbouring countries, their government, revenue, and manners.[4]

Alexander Hamilton (1757-1804)

Hamilton, appointed to join Bogle, was also of Scottish descent, although born on the West Indian island of Nevis. The two young envoys and Purangir Gosain, an agent of the sixth Panchen Lama, left Bengal in May 1774 for Bhutan. They remained there for three months, conducting diplomatic exchanges, before continuing towards Tibet via a pass twenty-four kilometres south-west of Chomolhari (and therefore well within sight of Kangchenjunga). The only map available to assist Hastings in planning Bogle's journey was d'Anville's *Nouvel Atlas de la Chine, de la Tartarie Chinoise et du Thibet*,

---

4 *Narratives of the Mission of George Bogle to Tibet and of the Journey of Thomas Manning to Lhasa* by Clements Markham, 1876.

published more than forty years before. Although the basic alignment of the mountains, rivers and lakes was drawn with reasonable accuracy, detail was thin. Bogle was later to confirm in his reports that the map was correct in its positioning of the Tsangpo river, at least where he had to cross it.

After passing below the mountain, while travelling along the Chumbi Valley, Bogle wrote in his journey notes:

> You see the high mountains, in the Demo Jong [Sikkim] country among which I imagine, is the snowy hill seen from Dinajpur and other plains in Bengal.

He can only be describing Kangchenjunga and is the first European so to do.

Julius Heinrich von Klaproch (1783-1835)

By then, other maps had been published of this area. The first was by Klaproch. It covered all the mountain areas of the Himalaya including those of the Karakoram and Sikkim, called here 'Djimoula'. Colonel Burrard, Superintendent of the Trigonometrical Survey, in his *A Sketch of the Geography and Geology of the Himalaya Mountains and Tibet* (1908) seems to have been impressed that Klaproch had positioned the mountain (Kangchenjunga) and town of Phari correctly.

Captain George Kinloch of Kinloch (1775-1833)

One unexpected opportunity to penetrate forbidden Nepal and enter the Himalayan foothills was taken by Captain Kinloch, seven years before Bogle and Hamilton's journeys into Bhutan and Tibet. Kinloch was despatched in 1767 to cross the fever-infested Terai of Nepal in support of the Malla rajah, who had requested British assistance in repelling an attack by the neighbouring Gurkhas under their leader, Prithvi Narayan Shah (1723-1775).

William Kirkpatrick (1754-1812)

Sent into the Kathmandu Valley to negotiate an agreement between the Chinese and the Gurkhas, Kirkpatrick was militarily assisted by Ensign John Gerrard. By the time the negotiators had arrived, the battles were over and the victorious Chinese had tightened their grip on Tibetan territory. Trade between India and Tibet became even less likely. This reverse was offset only by the gain of notes and sketches made of the route to the Kathmandu Valley (one of the principal access corridors to the Kangchenjunga massif) by Kirkpatrick and Gerrard.

Other British visitors had reached Nepal, but Kirkpatrick and Gerrard were the first to bring out information on the topography, the difficulties of travel, the history of Nepal and an inventory of traded goods.

Rishikesh Shaha has edited Kirkpatrick's report, which eventually came to light in Kirkpatrick's *An Account of the Kingdom of Nepaul*, published in 1811. His *Account* was by far the most comprehensive of its kind, certainly in the English language. It was a remarkable achievement, considering that he was only in the country for seven weeks and most of his information had come his way second-hand. His work was greatly enhanced by Ensign John Gerrard, whose recorded observations of the route to Kathmandu and surrounding peaks formed the first reasonable map of that part of Nepal.

In 1800 yet another political crisis developed in Nepal, when the rajah, Ranbahadur (1775–1806), was forced to seek sanctuary in British India. He was accommodated at Benares (Varanasi), to where many Nepali government officials who had fallen out of favour had previously fled. The Gurkhas in Kathmandu, fearing the British might exploit their hosting their former leader, agreed to discuss a new commercial treaty with a British envoy who could remain in the capital as British resident.

Captain William O. Knox (1789–1825)

Knox, who had taken part in Kirkpatrick's mission, was now sent to secure a treaty that would not strain relations with the Chinese. In 1803, Knox was withdrawn and the residency terminated. It had not all been a waste of time, however, since more information on the geography of the country and the mapping of Nepal had been gathered.

Dr Francis Buchanan
(sometimes Buchanan-Hamilton, 1762–1829)

Buchanan accompanied Captain Knox to Kathmandu in 1802. From 1802 to 1803 he spent fourteen months in the country, collecting information on all aspects of the Kathmandu Valley and beyond, resulting in *An Account of the Kingdom of Nepal and the Territories Annexed to this Dominion by the House of Gorkha* (1819).

Captain Charles Crawford (1760–1836)

At this time, Crawford was appointed commander of the Resident's escort. During the Residency, Crawford was (according to Buchanan's report to the Surveyor General of Bengal, Robert Colebrooke) able to settle:

the Longitude and Latitude of Catmandu by a numerous series of celestial Observations and has formed a Map of the Valley of Nepaul on a large scale constructed trigonometrically with great exactness and immense labour.

Crawford corrected the previous survey by Kirkpatrick and added to the map all the hills visible from the Kathmandu Valley. With information provided to him and Buchanan from local merchants, religious mendicants and other travellers, he was able to produce a small-scale map of the whole country. He was promoted to Major and encouraged later, in the winter of 1804–05, to survey that part of eastern Nepal where the Kosi emerges from the foothills north of Biratnagar, less than 160 kilometres south of the Himalayan divide and Kangchenjunga, on the Nepal–Sikkim border. Crawford again brought to the outside world's attention the great height of the Nepal Himalaya. In 1813 he was promoted to Colonel and appointed Surveyor General in recognition of his devotion to the accuracy of his cartography.

Major-General David Ochterlony (1758–1825)

In 1814, the Governor General, Lord Moira (later the Marquess of Hastings), declared war on Nepal. The only British general to distinguish himself, Major-General David Ochterlony, negotiated with the Kathmandu Durbar to make the Treaty of Sugauli, ratified on 4 March 1816. Sikkim, including Darjeeling, was completely liberated from Nepali occupation and rule, with the Mechi river becoming Nepal's new eastern border as far as the Terai. This treaty, apart from defining boundaries, also allowed for a British Resident to enter Kathmandu once again, although now the position was elevated from *Confidential Person* as in the 1801 Treaty to *Accredited Minister*.

Edward Gardener, from 1816 to 1829, became the first Resident, with Brian Hodgson as his Secretary from 1820. In total, Brian Hodgson came to live continuously in Nepal for twenty-three years, during which time he devoted himself to studying every branch of knowledge associated with Nepal and Tibet.

Brian Hodgson (1800 or 1801–1894)

Hodgson was a polymath of the English Enlightenment. He was born in Cheshire of a family who had fallen on hard times. He attended Haileybury, the East India Company college for educating administrators, as a pupil in the House run by Thomas Malthus. According to Hodgson's biographer,

W.W. Hunter, in *Life of Brian Houghton Hodgson, British Resident at the Court of Nepal* (1896), Hodgson, under the influence of Malthus, changed from a 'young aristocrat in social feelings and sympathies [into] an advanced liberal in politics'. He finished top of his year in Bengali, Persian, Hindi, political economy and the classics and took up a Company appointment in Calcutta. Hodgson's constitution suffered in the tropical climate, Calcutta being situated on a malaria-infested swamp. He left this centre of politics and a vibrant cultural life for the foothills of the Himalaya. After a year in Kumaon, Hodgson was appointed Assistant Resident in Kathmandu. He was fortunate to find himself working under Edward Gardener, a former Commissioner of Kumaon. He began to amass Sanskrit manuscripts until he was recalled to Calcutta as Deputy Secretary of the Persian department of the Foreign Office. Unfortunately for his career, he had to give up this promotion on health grounds and returned to Kathmandu in 1824, in the relatively humble position of Postmaster. In 1829, he was promoted to Resident and remained so until 1843, during which his life's work took place.

He delved deeply into the flora and fauna of Nepal, the character, habits and religious beliefs of the different ethnic people of Nepal, the architecture of the Kathmandu Valley and the country's geography.

The Resident was generally restricted in movement to no more than a morning's walk from Kathmandu. On two occasions he was allowed to make further excursions, once a few miles westwards to the banks of the Trisuli river and on another to the Sunkosi in the east. Despite these restrictions, he came to know more about Nepal than any other Westerner. He relied on information obtained from a variety of merchants and various mendicants whom he had befriended. These contacts helped him to put together his 1857 map, which clearly shows the position of Kangchenjunga (but not Everest): no small achievement.

Through his friendship with Jung Bahadur (1817–1877), ruler of Nepal and initiator of the Rana dynasty, Hodgson is also thought to have been instrumental in maintaining Nepal's neutrality as a buffer state. Jung Bahadur was to play a significant part, indirectly, in opening Kangchenjunga to the British, while on occasion giving vital direct support to the exploration of Kangchenjunga.

Before moving on to examine the exploration of Kangchenjunga itself and how it was climbed, it is worth looking at the character and circumstances of those who had been opening the way towards the world's third-highest mountain.

Brian Hodgson fits into this mould of British Indian civil servant. Although not a District Commissioner, he nevertheless showed in his twenty-three years in Nepal that he was tenacious, resourceful, respectful of custom and culture, conducted himself with integrity and remained at his post despite the lure of home. Maybe in his case, as with so many of his kind, he had severed his connection with home during his formative years. He never really suffered homesickness, never suffered the depression that can affect those who are forever longing for home when abroad. The explorers and climbers who climbed on Kangchenjunga, up to the first ascent in 1955, demonstrated this personality, with one or two notable exceptions.

4
Measuring the Heights

In 1846, with the completion of the 'Great Arc', the Survey of India was able to commence its primary triangulation along the foothills of the Himalaya, then beyond.

During work on the north-east 'Longitude North' series from Dehra Dun to Assam, surveyors became convinced from their theodolite readings that the Himalayan peaks were higher than anywhere else, higher even than Chimborazo (6,268m), long thought to be the highest point on the planet.

The Surveyor General of Bengal, Robert Colebrook, was interested in knowing more about the greatest heights. Charles Crawford (see Chapter 3) had surveyed the Nepal Valley as far as Kathmandu while escort commander for the first British Resident. In 1804, he also surveyed part of Eastern Nepal, and returned to announce the great height of the 'Snowy Range'. Colebrook encouraged his assistant, Lieutenant William Webb, to investigate further. From 1809 to 1810, Webb surveyed the position and heights of Dhaulagiri from four stations; the height was calculated at 26,862 feet (8,190m), only 67 feet (21m) below the height given today. In Britain, these measurements were derided as 'fanciful'. It was not until the Survey of India work under George Everest (pronounced 'Eve-rest') had been accomplished that the Himalayan range was accepted as the highest, despite Nanda Devi having been measured by James Herbert, the surveyor of the north-west provinces, in 1822, as 25,749 feet (7,850m).

Inevitably, Kangchenjunga would be among the first of the highest mountains to be accurately surveyed as it was one of the most prominent and accessible, situated only 560 kilometres north of Calcutta and 70 kilometres north of Darjeeling.

It fell to Everest's successor as Surveyor General, Major-General Andrew Scott Waugh (1810–1878), to complete the Gangetic grid of the north-east

'Longitudinal' series. It was concluded where the 'north-east' series intersected with the survey running north along the Calcutta meridian: Sonakhoda (on the arc between Calcutta and Darjeeling) has become an important place in the history of Kangchenjunga since the two series joined there. Waugh himself, on a clear winter's day in November 1847, measured the vertical and horizontal angles of Kangchenjunga's summit from there. Other readings by Waugh's assistants were later taken around Darjeeling, from viewpoints such as Tiger Hill, and further afield, from Senchal and Tonglu. On the map accompanying Col. A.S. Waugh's paper *On Mounts Everest and Deodanga*, published by the Royal Geographical Society in 1858, readings were taken from six stations on the plains in the vicinity of Sonakhoda.

The height calculation was only announced by Waugh in 1849 in a Survey departmental memorandum, which declared:

> the Western Peak of Kangchenjunga obtains an elevation of no less than 28,176 feet above the sea, which far exceeds what has hitherto been conjectured.

There was no hint of triumphalism about recording the highest peak ever measured. John Keay, in his comprehensive *The Great Arc* (2000), suggests that not only was Waugh far less flamboyant and demonstrative than Everest, but he was also just as thorough and as likely to follow the Survey of India procedures as Everest did. Neither Everest nor Waugh, however, considered surveying the Himalayan heights as much more than of passing interest.

Waugh, in 1849, released the 'fact' that Kangchenjunga was the highest point known to man. This remained the position until 1856, by which time all the measurements on peak 'Gamma'/'b'/'XV' had been computed, giving a figure of 29,002 feet (8,842m). Kangchenjunga was demoted to second-highest peak; soon it was to go down to third. Twelve hundred kilometres away, Thomas Montgomerie, working on the Kashmir Survey, had first scoped a high mountain which he labelled 'K2'. Two years later, George Shelverton fixed the height at 28,287 feet (8,624m) and in August 1861, in the first close-up measurements, from a point on Masherbrum about 1,000m above Urdukas on the Baltoro glacier, Haversham Godwin-Austen fixed K2's height and coordinates with remarkable precision. Thus, in 1861, Kangchenjunga slipped back to third highest, K2 to second and Peak XV, later 'Everest', was universally accepted as pre-eminent.

Kangchenjunga will always enjoy the accolade of being the first of the

8,000m peaks to be accurately measured by the Great Trigonometrical Survey, begun at St Thomas' Mount, 1,900 kilometres away in the south-east of India. The significance of a religious conversion connection is highlighted in *Khangchendzonga: Sacred Summit* (Wangchuk and Zulca, 2007): St Thomas is believed to have brought Christianity to India. Seven hundred years later, Padmasambhava, the Second Buddha, travelled through Sikkim, subduing harmful spirits and converting the Sikkimese and Tibetans to Buddhism. He appointed dZanga (*sic*), who resides on Kangchenjunga, the 'Supreme Guardian' deity. Padmasambhava hid religious texts, known as *ters*, around Sikkim. They would, at certain times in the future, be rediscovered by *tertöns*.[1] All this confirmed that Sikkim was the most holy of all the hidden lands and that Kangchenjunga was its supreme guardian deity.

Kangchenjunga: the British connection

During the nineteenth century, India was becoming ever more widely controlled by the British. In 1835, the British had commandeered the small ridge-top settlement of Darjeeling, ostensibly as a sanatorium above the Gangetic plains for convalescing East India Company staff and army officers.

The town occupies a 2,043m hilltop south of the Great Rangit river, which flows west to east from the Singalila Ridge down a steep, 1,500m-deep ravine to join the Tista, the main river of Sikkim and northern Bengal. Looking north, the land drops away to give a commanding view across the foothills to Kangchenjunga, seventy kilometres distant. This famous, much admired view was considered at the time more captivating than anywhere else in the Himalaya (see Chapter 6).

By 1840, a road connecting Darjeeling to the plains had been constructed. Very quickly thereafter, Darjeeling became a well-laid-out and rapidly growing hill station. The town was not only popular with British officials and army officers, but also attracted Sikkimese citizens, especially labourers, the occasional runaway criminal, and slaves, all anxious to escape the Sikkimese authorities by becoming a British subject. This did not go down well with the Sikkimese government, who were to cause problems for Joseph Hooker and Archibald Campbell on their journeys into Sikkim and up to the Tibetan border.

---

1 *Tertön*: Tibetan for a person designated to rediscover ancient and/or holy texts.

The surveyors of Kangchenjunga and its environs
Captain Walter Stanhope Sherwill (1815-1890)

Captain Sherwill – not to be confused with James Lind Sherwill, who came to the Kangchenjunga region as a mountain tourist ten years later – first went to India to work for the East India Company in 1834. He served as an ensign and then as Revenue Surveyor for the Bengal Presidency until his retirement in 1861, when he was gazetted the 'Honourable Lieutenant-Colonel'. In 1852, as Revenue Surveyor and on geological fieldwork, he visited Sikkim and then, via the Kalet Chu, crossed the Singalila Range to examine the geology of the north-west side of Kangchenjunga and 'the perpetually snow-covered peaks in its vicinity'. His findings were published in the *Journal of the Asiatic Society of Bengal* in 1854 as 'Notes of a journey in the Sikkim Himalayas'. He had surveyed and mapped as far as the source of the Yalung, Yunga and Ringbi rivers. Hooker (see Chapter 5), having recognised Sherwill's greater accuracy, incorporated this survey into his map. In 1852, Sherwill eventually produced a manuscript map of British Sikkim including 'Darjiling Hill Territory', reduced in 1853 to a scale of four inches to one mile and eventually published in Calcutta.

Captain Henry John Harman (c.1830-c.1883)

Captain Harman, of the Survey of India, was the next exploratory cartographer of Sikkim after Hooker in 1850. He added to Hooker's map after surveying the route from Darjeeling to Tumlong during the Gawler military expedition of 1861.

Captain Harman had worked on the Great Trigonometrical Survey from 1872, soon after his arrival in India. Following active service in the Garhwal and Kumaon, then from 1874 until 1878 in the north-east of India, in 1878 he moved to Darjeeling where he was given responsibility for the mapping of Sikkim up to its boundary with Nepal. Harman made several journeys from 1879, with varying success. At his first attempt, he was forced back from reaching the monastery of Tulung by local hostility.

In November 1879, he reached the Donkia pass in cloudless weather and ideal conditions for observing the surrounding peaks. He arrived very late in the evening and decided to bivouac on the pass, at 4,250m, with his two Tibetan assistants. They had only one blanket between them, so they lit a fire, but Harman suffered severe frostbite in the feet, losing several toes. He returned to Darjeeling where he seemed to recover his general fitness and set about trying to solve the riddle of the course of the Tsangpo–Brahmaputra.

The pundit explorations of Kangchenjunga and its surroundings

It was Thomas George Montgomerie (1830–1878) of the Kashmir Survey who formed a favourable impression of native surveyors, his *khalasis* (manual workers), without whom there would have been no Survey of India.

By 1865, it was becoming necessary to survey the forbidden lands to the north of British India, partly to understand in what ways the Government of India could rely upon the Himalaya as a defensive barrier; partly to explore the trading opportunities with Tibet and China; and partly to know the geography, the culture and customs of Tibet, Nepal, Sikkim and Bhutan.

By 1878, plans were being made for three *pundits*, local to Bengal and Sikkim, to explore the various passes through the Himalaya to Tibet and round the north side of Kangchenjunga between Nepal and Sikkim.[2] All three had a strong connection with the Bhutia Boarding School at Darjeeling.

Sarat Chandra Das (1849-1917)

The Bhutia Boarding School was opened in April 1874; the first headmaster was Sarat Chandra Das ('Pundit S.C.D.'). He was born in Bengal in the year the British annexed the Punjab, extending their Indian Empire to the turbulent borders with Afghanistan and Kashmir and thus ensuring future work for the pundits: Russian troops were then only 1,000 kilometres from the Indian border. The school was intended by Sir Andrew Croft, the Director of Public Instruction in Calcutta, to:

> train interpreters, geographers and explorers who may be useful at any future time that Tibet might be opened to the British.

Sir Andrew had been a mentor of Chandra Das; Das was naturally chosen to head the school. Lama Ugyen Gyatso, one of the 108 monks at the Pemayangste monastery, taught there. He became a pundit under the recorded name 'the Lama' or 'U.G.'

Chandra Das visited Pemayangste monastery, knowing it to be Sikkim's most ancient and important seat of Buddhist teachings and the place where the *chogyals* (Sikkimese rulers) are traditionally enthroned. From the monastery, prominently sited on a hilltop, splendid views of Kangchenjunga, forty kilometres to the north, open out. There, in 1876, Das began to dream of visiting Tibet, an urge strengthened later after reading Sir Clements

---

2   *Pundit*: from the Hindu *pandit*, a learned person or expert.

Markham's newly published *Narratives of the Mission of George Bogle to Tibet and of the Journey of Thomas Manning to Lhasa* (1876).

The Deputy Commissioner at Darjeeling, Major Herbert Lewin, opposed the idea, so Das thought up a plan that might appeal to the British authorities without the approval of the Bengal government: he was to have his assistant schoolmaster, Ugyen Gyatso, being a lama of Sikkim's most important monastery, visit Lhasa and Tashi Lhunpo monastery with appropriate tribute. At the same time, he would try to arrange for an official invitation and passport for Das to visit Tibet. Das gained the tacit support of Sir Andrew Croft, although he remained sceptical of any success.

Ugyen Gyatso set off for Tibet in May 1878 and returned in September with a passport issued by the Tashi (Panchen) Lama for both pundits. Das and Ugyen Gyatso, having both been replaced at the school, were despatched to Calcutta where, under the direction of surveyors Tanner and Harman, they were taught surveying methods and shown the techniques of clandestine exploration and mapping.

Much information about their odyssey comes from Das's narrative *How I crossed the Jon-Tsang La Pass*, reprinted in Appendix C of Douglas Freshfield's *Round Kangchenjunga* (1903), a useful addition to this masterpiece of mountain literature.

The 'expedition' left Darjeeling in June 1879 with a guide, Paljor, and two porters, reaching Jongri (Dzongri) on 17 June. From there, they headed north-west to the Nepali frontier. Their passport could not have been more helpful in being wide-ranging, giving Sri Sarat Chandra Das and his retinue a choice of routes. In the document were instructions to the *dzongpön* (district magistrates) at all the borders to facilitate the progress of the travellers and their luggage.

On 19 June they crossed the Rathong by a bridge of planks and continued towards the Nepali frontier through 'endless groves of rhododendron'. Lama Ugyen Gyatso's task on the journey was to take bearings for the route survey, while Das was to plot their exact position by making frequent sextant observations of the stars. In the pre-monsoon season, neither found their tasks easy. On 20 June, after camping out under rocks and in caves, they climbed on snow:

> The lama and I put on our blue spectacles, while our coolies and guides painted their cheek-bones below the lower eyelid with black to protect their eyes from the glare.

After well over a kilometre on the snow, they came to rocky ground where flags were flying. The guide announced they were about to descend into Nepal. Das recorded the geological changes:

> All the rocks and boulders on this side of the Kang-la were of red sandstone while in Sikkim most of the rocks are of siliceous, calcareous or granitic formation.

They then crossed the Kang La pass (c.5,054m). By 22 June they had descended to the Yalung river, which they crossed by a bridge of deal (a type of fir) planks and juniper logs, then began to cross one pass after another to Ghunsa. They were following an ancient caravan route which connects the Yalung Chu (river) with the Ghunsa Khola (river). Their Western predecessors for this crossing were Joseph Hooker in 1848 (see Chapter 5), followed by Elizabeth (Nina) Mazuchelli's party in 1872 (see Chapter 7).

These passes were known collectively as 'Chunjerma', after the local name for this mountainous country. 'Choonjerma' is Hooker's spelling; he depicts it in a fine watercolour with deep snow lying on it, with Jannu the dominant peak to the north.

The various passes are known today, beginning with Tseram (3,870m) in the Yalung Valley (the river is marked on most trekking maps as the Simbua Khola), as the Sinelapche La (4,712m), Mirgin La (4,689m), Sinion La (4,661m), Sele La (4,116m) and last the Taka La (3,981m), which drops to the Yamatari Khola, a feeder of the Ghunsa Khola and Ghunsa village at 3,540m. Kev Reynolds comments in *Kangchenjunga: A Trekkers' Guide* (1999):

> Confusion reigns as to which La is which. Some authorities claim that the first is the Mirgin La, the second being the Sinion La. Others argue that the Sinion La is the first to be crossed, while Bezruchka also names this Menda Puja. The Ghunsa man who accompanied my last crossing called the first La the Sinion La, the second the Mirgin La and said that the third, unnamed pass was called the Deorali Danda – but this is a common Nepali name which merely indicated the ridge-top of a hill. (In 1881, Chandra Das called it the Nango Lap-tse.) As if this confusion were insufficient, Hooker referred to the Mirgin La as the Choonjerma Pass – but, according to Chandra Das, Choonjerma (meaning 'collection of cascades') is the rocky terrace between the Mirgin La and the unnamed pass.

Chandra Das also refers to yet another pass between the 'Seenon-La and Mirkan-La' as the Pangbo La. All very confusing!

The two pundits continued along this ancient trading route between Sikkim (Pemayangtse Gompa/Yoksum) and Tibet (via Tiptala or Yangma and the Ghang La). They set off early and by noon they 'reached the top, where there are two small lakes'. After crossing four ridges:

> at 6 p.m. we reached the beautiful village of Kambachen-Gyunsar (9,500 feet) which is situated in a romantic valley on the banks of a fine river and overhung on three sides by steep and rugged mountains, covered with thick woods of rhododendron, juniper, deodar and weeping willows.

They were well received by the friendly villagers, with Das able to pass himself off as a lama of Nepal. They crossed the Ghunsa Khola to the Tashi-Choding monastery with its 'eighty monks, besides a dozen nuns who generally reside in the village. The monastery is one of the finest and richest in Sikkim and Eastern Nepal.'

The lamas here, and at the Pemayangste monastery in Sikkim, belong to the same Nyingmapa or 'Red Hat' sect. The pundits received an exceptionally warm welcome. The villagers and lamas provided the strongest man in the village, a monk called Phurchung, to serve as guide to the party, now with new 'coolies' in place of those who had reached Tashi-Choding before returning.

At 7 a.m. they set off, to arrive at a barley mill and a long *mendong* of *mani* (prayer) stones marking the entrance to Kambachen.[3] They were very impressed by the wooden houses in such a beautiful setting. Here and at Ghunsa, Das noticed that:

> No nails or ropes are used to fasten the planks to the rafters or to each other, but they are kept in their places by blocks of stone laid on them.

They witnessed:

> the grand offering made to the Kang-chan peak by the residents of Gyunsar and Kambachen. The firing of guns, athletic feats,

---

3   *Mendong*: in Tibet, a drystone wall in which flat stones bear sacred texts or images.

and exercises with the bow and arrow form the principal parts of the ceremony, which is believed to be highly acceptable to that mountain-deity. The youth of Gyunsar vied with each other in athletic exercises; the favourite amusements of their elders being quoits, back-kicking, and the shooting of arrows. We also contributed our share to these religious observances. The scene reminded one of the Olympic games; and like good Buddhists, we too paid our obeisance to Kang-chan, the Indian Olympus.

They left hurriedly, having been warned by the monks and village headman that the Tibetans were about to close the passes to all traders with yaks and sheep, owing to cattle disease.

At dawn on 26 June, they set off, after five kilometres passing a waterfall called *Khan-dum-chu*, a very holy place where eight Indian saints are said to have bathed. Das estimated the final fall to be 300 metres high and, where it emptied into the Ghunsa Khola, six metres wide. They continued upstream by mainly level valleys, the quiet beauty of which contrasted with the sublime beauty of the surrounding hills, until at noon they reached Ramthang, where they ate their breakfast in a yak shed. Nowadays, on the walk from Kambachen, Pangpema Ramthang (4,616m) is the name given to a *kharka* or pasture opposite the moraine of the Ramthang glacier.

(The third pundit to visit this area, Rinzin (Rinzing) Namgyal, in 1883–84, marked Ramthang as 'Lhonak'. Freshfield, in his account of the circuit of Kangchenjunga, reverted to using 'Ramthang', which was subsequently placed on Garfield's map because he thought Lhonak would be confusing, it being the name used for north-east Sikkim.)

A footnote in Freshfield's narrative states, 'From this point the Pundits' route seems to diverge from that of the Jonsong La.' From there on, Das's description suggests that they crossed into Tibet via a pass well to the west of the Jonsong La.

Between 27 and 30 June, according to Das's narrative, the pundits and their lama guide, Phurchung, with the 'coolies', crossed the Chathang La (named alternatively as the Jon-Sang La) then dropped to a river which descends to the headwaters of the Teesta. On the following day, they climbed to the Chorten Nyima La, then marched into Tibet.

Das did not acclimatise well, which may account for there being very little information to help the reader know where his party went from Lhonak onwards. Freshfield comments in *Round Kangchenjunga* (1903):

I permit myself to doubt whether the pass he traversed ... in 1879 to Tashilumpo in Tibet was identical with our Jonsong La ... In his original printed narrative, he called his pass the Chathang La ... Das seems not to have applied the name Jonsong La until after one of his native colleagues in the Survey had crossed this latter pass ... Das's published sketch-map is, unfortunately, in some parts too vague, in others too inaccurate, to be intelligible. But the details of his narrative, if any trust is to be placed in his compass and the direction he assigns to the surrounding ranges, furnish conclusive evidence that his pass and ours were not the same. It is further to be noted that the late Colonel Tanner tells us [*Survey Report*, 1883–84, p.7] that the Pundit's observations 'place his pass considerably west of the Jonsong La'.

Colonel Tanner concludes that the Babu's observations were at fault. But it is surely at least as plausible to suppose that the identification of Chandra Das's pass with the Jonsong La was erroneous. There is, however, I think no doubt that the 'Chortenima La' crossed by Chandra Das was the pass known by that name in the country. The sketch he gives of the strange crags on the top corresponds very fairly with Mr White's photograph taken in 1892.

Freshfield goes on to say that whether Das 'crossed one high pass or another makes not the smallest difference', as it was a marvellous feat to have negotiated the dangers of snowfields at great height, especially as he was 'an Indian born on the shores of the Bay of Bengal'.

Freshfield implies Das's party crossed into Tibet by the Chabuk La (6,152m) at the head of the Lhonak glacier. That would then involve a considerable journey west to east along the north side of the main Himalayan divide, before crossing the continuation of the Singalila Range north of Jonsong Peak to reach the South Lhonak glacier. From there, a descent would be needed to pick up the trail leading to the Chorten Nyima La, a walking distance of thirty kilometres or more from pass to pass.

From Lhonak, the alternative route could conceivably have followed the east side of the Lhonak glacier, via either the Tsisima or even the Broken glacier, but a study of the very detailed Swiss 1:150,000 *Sikkim Himalaya* map indicates the Chabuk La would involve less steep ascent and descent, despite being shorter in distance. Whichever way they went – and, of course, they may have gone over the Jonsong La – it was not easy.

Das came to regret it as 'imprudent and ill-judged' that he had chosen, too late, a rarely followed route into Tibet. The pundits were suffering not only from the altitude but also from lack of fuel to cook their rice properly. Being undernourished, they struggled to wade through soft snowfields at around 6,000m. In fact, Das was carried by the indefatigable Phurchung:

> In this miserable fashion did I cross the famous Chathang La [Jonsong La] into Tibet, the very picture of desolation, horror and death, escaping the treacherous crevasses which abound in this dreadful region.

Against Das's expectations, Lama Ugyen Gyatso, despite being corpulent, managed to cope with the height and snow conditions on his own.

They eventually completed their journey to Shigatse and the fabulous Tashi Lhunpo monastery, 'like a dazzling hill of polished gold', by 7 July. After teaching the chief minister some Hindi and Sanskrit they returned to Darjeeling in late September via the Donkia pass, with forty volumes of Tibetan manuscripts and an invitation to return the following year. The British Government of India was delighted to acquire new knowledge of the topography of north-east Nepal; even more important was the data on customs and culture of Tibet, including an indication of trading opportunities – and all at no cost to the public exchequer.

Owing to the turbulence in Sikkim concerning the immigration problem, Das had to postpone his second visit until 1881. On 7 November, Das, the lama and their faithful guide, Purchung, left Darjeeling for a second visit to Tibet. They again crossed into Nepal, this time via the Chumbok La, south of Kang La Peak, followed by the Semo La, probably as Freshfield indicates, the way taken earlier by Elizabeth Mazuchelli's party.

They again visited Ghunsa, where they exchanged 'coolies' for local porters who, suitably laden with food and blankets, headed for the Nango La, which Hooker had been the first European to cross. In general, Das followed Hooker's route to Yangma and over the Kanglachen La.

Again, Das's account is vague but, generally, his route lay much further west than before, following an eastern tributary valley of the Arun to Tashirak village in Tibet, where he left the upper basin of the Arun catchment by way of the Langbu La (5,200m).

Das did continue to be employed in secret, 'on political duties in the direction of Tibet'. When Das retired from exploration, his friend, Lama Ugyen Gyatso, and the lama's brother-in-law, Rinzin Namgyal (Pundit

'N.R.') continued to explore the environs of Kangchenjunga, with Rinzin completing the first circuit of the massif.

Lama Ugyen Gyatso (1851–c.1915)

Ugyen Gyatso was by no means the silent companion on his journeys with Das, as his task was to act as secretary, and to deal with local people. He was the expedition's translator, while also making a survey of the topography (despite not being very well trained). The lama certainly complemented Das's work: between them, more was revealed of the vegetation, geology and human landscape around Kangchenjunga than ever before.

Ugyen Gyatso was born into a distinguished Sikkimese–Tibetan family whose ancestors in AD 1073 had established the Sakya monastery in southern Tibet and later the Samduk Lhakhang (temple) in the upper Chumbi Valley town of Phari. His great-uncle had founded a monastery at Yanang in central Sikkim, the place of Ugyen Gyatso's birth. With these antecedents he was clearly destined for holy orders. He entered the Pemayangtse monastery when he was ten and remained there for twelve years, to complete his training as a lama.

When twenty-one, the lama was sent by his monastery to bring back from Tibet a complete set of the Buddhist scriptures, known as the *tongyu*. This operation required a whole year, after which Gyatso visited Darjeeling, then journeyed to the Sikkim and Tibetan border in the company of the Deputy Commissioner, J.W. Edgar, an acquaintanceship which secured his selection in 1874 as teacher at the Bhutia Boarding School in Darjeeling.

He was formally listed as an agent of the Survey under the pundit initials 'U.G.' His cover was as schoolmaster at the Darjeeling Tibetan school. Colonel H.C.B. Tanner of the Survey of India had 'U.G.' attend his basic training course for surveying as a spy. He was politically instructed by a British administrator, Colman Macaulay.

'U.G.' and his party entered the Chumbi Valley near Phari but bypassed it at night for fear of recognition. They only stopped at the town of Chumbi, where they had relatives who celebrated their safe return with a feast in an outlying cave. 'U.G.' completed his 'long and diligent pace-and-compass route survey' on 17 November when his party crossed from Tibet into Sikkim via the Cho La. They returned to Pemayangtse monastery on 7 December and Darjeeling on 15 December. This third journey as an agent of the Raj was 'U.G.'s most remarkable and, according to Colonel [later Sir Thomas] Holdich of the Great Trigonometrical Survey, 'one of the best records of Tibetan travel that has yet been achieved by an agent'. (Gyatso's topographical work would be invaluable to the Younghusband mission during the British invasion of Tibet in 1904.)

After his active exploring came to an end, 'U.G.'s service was properly recognised by the Indian government: in 1893 he was presented with their silver medal and, from the Viceroy, a letter which described his achievements. The award was for:

> distinguished services rendered by you in obtaining geographical and statistical information in the little-known regions of the North East Frontier of India during explorations carried out at great personal risk and hardship.

He was a most unlikely explorer, considering his corpulent build and penchant for getting into apparently impossibly difficult situations. His character and confidence, however, were such that he was always able to extricate himself. As no financial gain was involved, and he was not seeking kudos from his own countrymen for facing up to so many hardships, it can be assumed his driving force was the pursuit of knowledge and adventure for its own sake.

Rinzin (Rinzing) Namgyal (1850-sometime after 1902)

Rinzin Namgyal was the brother-in-law of Lama Ugyen Gyatso ('U.G.') but there is no indication that it was the same brother-in-law who travelled round south-east Tibet with 'U.G.' and his wife. Rinzin became the first person to make a complete circuit of Kangchenjunga; he greatly enhanced the mapping of the massif.

He was born into an aristocratic Sikkimese lama family and attended the Bhutia Boarding School in Darjeeling. Under the direction of Colonel Tanner, his first important survey, of the Talung Valley, took place in 1883. No European had been there previously. This survey put in place one of the last sets of information to complete the map of Sikkim.

One blank on its map remained: the border area between Nepal, Sikkim and Tibet to the north of Kangchenjunga. By 1884, the new Deputy Superintendent of the Survey of India, Colonel Tanner, was convinced that his protégé now possessed all the qualities and experience for carrying out a survey round the Kangchenjunga massif, to map the valleys and spurs descending into north-east Nepal. 'R.N.' entered Nepal in the middle of October 1884, to reach the Yalung river on 19 October. They spent time surveying the upper Yalung, reaching Yalung Kang 'Snowy Peak', at about

5,800m. 'R.N.' returned to Tseram, leaving this collection of nomads' huts on 2 November, to arrive at Ghunsa the following day. Once again, they were impressed by the hospitality and lifestyle of the inhabitants who:

> are well-to-do people generally engaged in trading business. The women here spend their time in weaving blankets. Men and women in the village every night go from one family to another to interchange visits, where they are treated with courtesy and presented with cups of *mowa* (a kind of liquor) and fruits, etc. Thus, they pass their nights in song and dance. In case of a death occurring in the village their jolliness is stopped for three days. They observe Tibetan customs. They keep yaks and goats and sheep. They are ruled by a headman, whose duty is to collect revenue and taxes. The village lands are fenced round to protect them against musk-deer and *burrel* and *munal*, which the villagers are forbidden to shoot.[4] The Gompa (monastery), being the repository of religious books and images, is governed by a Lama, whose supremacy is acknowledged by the villagers who monthly send food for him and his disciples (*dabas*). We passed five nights here.

On 9 November they reached the *pakka* houses, i.e. those designed to be solid and permanent, of Kambachen village. They were:

> found empty on account of the cold season and [were] surrounded by barley fields. To the east and opposite the village the Jannu (or Junnoo) Snowy Peak is visible, which is an object of worship by the inhabitants of Gunsa. There is a scarcity of firewood and grass. Here a companion of ours fell ill, and from this place he had to be carried.

After leaving Kambachen they walked for twelve kilometres to Lhanok, a cattle-shed, where a road diverges via the Chabuk pass to Tinkijong.

From this point, Sarat Chandra Das and Lama Ugyen Gyatso had probably continued north. Rinzin, however, now followed the Kangchenjunga glacier, covered mostly with moraines, east and then north-east, until on 16 November, after a final seven-kilometre glacier ascent, his party reached the Jonsong pass:

---

4   *Burrel* (sometimes *bharal* and other spellings): the Himalayan blue sheep.
       Munal (sometimes *monal*): the Himalayan pheasant.

on the seventh day from Gunsa ... this pass cannot be crossed unless assisted by some fifteen men in making a passage over snow. The nearest pass to the west is Chabukla, about 19,000 feet, and to the north Chhorten Nyima, which we afterwards visited; to the south the range cannot be crossed.

Colonel Tanner provides an interesting footnote at this point, relating to the previous visitors to this region and in particular Lama Ugyen Gyatso, who had been put in charge of the surveying on Das's expedition:

At that time the Lama was ignorant of surveying, and besides, the journey was made under the most trying circumstances, when snow fell daily, so that, even had he known how to take them, observations could hardly have been made. The route, however, was plotted and the map published, but the errors of the Lama's distances and angles were sufficient to throw the position of the Jonsong and Chhorten Nyima Passes many miles to the west. The Lama has crossed and measured the height of over ninety passes, and he informs me that he thinks the Jonsong is the highest and most difficult he ever attempted. In the first edition of the sketch map of North Sikkim by Mr Robert the name Jonsong has been placed opposite a wrong pass, from erroneous information supplied to that gentleman by his guides. A second edition of this sketch map will contain a number of corrections, and all the new geography to the north-west of Kinchinjunga by Rinzin Nimgyal.

Despite carrying one man over the pass in the depths of winter at great height, they had the drive to descend into Sikkim then climb all the way to the Chorten Nyima pass on the Tibetan border. It was a remarkable achievement to have moved this expedition over the pass (5,819m) without any formal training in mountaineering, on empty stomachs and in deep snow.

On the verge of starvation, they came across the tracks of two wild yaks, but after seven kilometres of following their prints they gave up. After their unsuccessful attempt to feed themselves, two of the party collapsed and died. Despite another four days without food, the remainder of the party reached Zemu settlement at the junction of the Lachen and Zemu rivers, where they accepted food from yak herders camped there.

They left:

> Zemusamdong cattle shed on the 5th December 1884, and marching for four miles to the south we reached Lachen or Lomting village ... We stopped here four days ... Having taken leave of the Snowy Range we came to Darjeeling via Cheuntong and Tumlong on the 31st January 1885.

Not only was this a great feat of endurance but 'R.N.' had succeeded in defining the frontier between north-east Nepal and north-west Sikkim as well as the position of the Jonsong La and surrounding peaks. The Surveyor General also noted that in crossing from the drainage of the Tamur River to the Tista:

> the explorers met with many glaciers in the northern valleys, and this contradicts the statement as to the non-existence of these glaciers made in Colonel Tanner's report of last year which was evidently based on erroneous information.

'R.N.' was now a highly respected pundit. He worked with Colonel Tanner at surveying Nepal from the GTS stations situated on the Gangetic plains before becoming a member of Freshfield's Kangchenjunga expedition of 1899 (see Chapter 7).

5

The Expeditions of J.D. Hooker

Samuel Turner's *An Account of an Embassy to the Court of the Teshoo Lama in Tibet* was published in London in 1800, just before his death. In spite of the hardships described, it fired the imagination of Sir Joseph Dalton Hooker (1817–1911), a friend of Darwin, a botanist of great distinction and soon to be the renowned first explorer of Kangchenjunga. Harish Kapadia makes an interesting comment in his *Into the Untravelled Himalaya* (2005):

> Though Kangchenjunga is the presiding deity for the people of Sikkim, to the British it was an object of curiosity and exploration. With their sense of enquiry in all matters, British explorers were attracted to Kangchenjunga as a mountain ... The first explorer to arrive in the area was J.D. Hooker, one of the greatest British botanists ... Why was he really there and how did he fit into the scheme of things?

Joseph Hooker was the first of an adventurous band of followers fascinated by Kangchenjunga, its land and its people. In many ways, Hooker's groundbreaking visit was by far the most important. His earliest reading recollections were of Turner's *Travels in Tibet* and of Cook's *Voyages*:

> The account of Lama Worship and Chumulari in the one, and of Kerguelen's Land in the other, always took a strong hold on my fancy. It is therefore singular, that Kerguelen's Land should have been the first strange country I ever visited ... At a later period I have nearly been the first European who has approached Chumulari since Turner's embassy. The possibility of visiting Tibet and of ascertaining particulars respecting the great mountain Chumulari

which was only known from Turner's accounts, were additional inducement to a student of physical geography [to explore Sikkim], but it was not then known that Kinchinjunga, the highest known mountain on the globe, was situated on my route.

In 1848, when Hooker arrived, all the valleys and glaciers descending from Kangchenjunga were unexplored by Europeans. Even to approach these valleys was difficult: it meant cutting through subtropical, roadless jungle with only occasional, vague trails. To negotiate steep jungle-covered hillside in the Eastern Himalaya, the indigenous hunter-gatherers have seldom bothered to cut tracks: footpaths zigzagging up the hillsides are rare. The local method of negotiating steep ground, by lashing together bundles of long bamboo to the stouter jungle vegetation, one above the other until there is a continuous line of ascent, is reminiscent of the British steeplejack edging his way up the side of a high-rise chimney by pulling up ladders to fix one above the other. None was more nimble on these bamboo structures, it seems, than the indigenous Lepcha who assisted Hooker with his plant-collecting.

Hooker and his party experienced poor weather during his explorations beyond Darjeeling. Early in his visit, he made a trip with local Resident and host Charles Barnes to Tonglo (Tonglu) Mountain (3,070m) on the 'Singalelah' (Singalila) Ridge, nineteen kilometres west as the crow flies but stretching to fully fifty kilometres on foot. Hooker and Barnes eventually reached the summit in bad weather and began to descend. On 22 May 1848, Hooker wrote:

> During the night the rain did not abate, and the tent roof leaked in such torrents that we had to throw pieces of wax-cloth over our shoulders as we lay in bed. There was no improvement whatever in the weather the following morning. Two of the Hindoos crawled into the tent during the night attacked with fever and ague. The tent being too sodden to be carried, we had to remain where we were, but with abundance of botany around, I found no difficulty in getting through the day.

What must strike anyone reading about Himalayan exploration is the high physical and mental calibre of the participants, none more so than Joseph Hooker. Such explorers were physically strong, courageous, very determined and exceptionally self-reliant, able to travel for months on end with only local people to places which foreigners had rarely, if ever, previously visited.

On returning to Darjeeling in June 1848, Hooker now took up residence with Brian Hodgson in a sprawling bungalow. With the monsoon beginning, they worked together as kindred spirits, discussing, writing, cataloguing the natural history of this small portion of the Himalaya that offered such an amazing diversity of flora and fauna. These two enthusiasts enjoyed each other's company and made huge progress in describing the ecology of the area within eighty kilometres of Kangchenjunga.

Hooker had been asked by Darwin and the distinguished geographer and polymath Alexander von Humboldt (1769–1859) to inform them about plant and animal life. Humboldt specifically requested temperature readings all the way to the Tibetan plateau. Unfortunately, Hodgson (who was frequently in poor health) had to withdraw from Hooker's expedition into East Nepal, although he contributed much to the success of this first penetration by a Westerner to the west of Kangchenjunga. Hooker recollected that:

> I owe it entirely to his [Hodgson's] personal influence with the late Sir Jang Bahadur [the effective ruler of Nepal] that I was permitted in 1848–49 to travel in East Nepal over ground never before or since traversed by any European and to visit the jealously guarded passes of the Nepali–Tibetan frontier.

Hodgson nevertheless did everything possible to make the expedition a success, as did Dr Archibald Campbell (1805–1874), whom Hooker acknowledged frequently in his journals. Hooker was highly dependent on his friends in Darjeeling for food supplies, as the Sikkim authorities remained obstructive.

The first expedition

On 27 October 1848, Hooker set off to visit north-east Nepal. On this expedition, he also crossed into Sikkim and explored the south-east side of Kangchenjunga. This first expedition was to occupy three months, during which he relied wholly on his party of fifty-six, as he explains:

> [the party] consisted of myself, and one personal servant, a Portuguese half-caste, who undertook all offices, and spared me the usual train of Hindoo and Mahometan servants. My tent and equipment (for which I was gratefully indebted to Mr Hodgson), instruments, bed, box of clothes, books and papers, required a

man for each. Seven more carried my papers for drying plants and other scientific stores. The Nepali guard had two coolies of their own. My interpreter, the Coolie-Sirdar or Head-man, and my chief plant collector, a Lepcha, had a man each. Mr Hodgson's bird and animal shooter, collector, and stuffer, with their ammunition and indispensables, had four more; there were besides three Lepcha lads to climb trees and change plant-papers, who had long been in my service in that capacity; and the party was completed by fourteen 'Bhotan' coolies laden with food, consisting chiefly of rice with ghee, oil, capsicums, salt and flour.

I carried myself a small barometer, a large knife and digger for plants, notebook, telescope and other instruments; whilst two or three Lepcha lads who accompanied me as satellites, carried a botanising box, thermometers, sextant and artificial horizon, measuring-tape, azimuth compass and stand, geological hammer, bottles and boxes for insects, sketch book, etc., arranged in compartments of strong canvas bags. The Nepal officer always kept near me with one of his men rendering innumerable little services. Other sepoys were distributed amongst the remainder of the party; one went ahead to prepare camping ground and one brought up the rear.

The 'Bhotan' men proved to be most unreliable:

> the contrast between the conduct of the 'Bhotan' men and that of the Lepcha and Nepalis was so marked that I seriously debated the propriety of sending the former back to Dorjiling but yielded to the remonstrances of their Sirdar and the Nepal guard, who represented the great difficulty we would have in replacing them.

By 3 November Hooker had returned to Tonglo in fine winter weather. He enjoyed the views that had escaped him during rainy May:

> In the early morning the transparency of the atmosphere renders this view one of astonishing grandeur. Kinchinjunga bore nearly due north, a dazzling mass of snowy peaks interced by blue glaciers which gleamed in the slanting rays of the rising sun like aquamarines set in frosted silver.

From Tonglo, Kangchenjunga of course dominated; but, to the north-east, Chomolhari (7,326m), which had inspired Samuel Turner and had later been surveyed by Colonel Waugh during November 1847, also stood out. Waugh and his Survey colleagues had realised that:

> the Western Peak of Kangchenjunga attains an elevation of no less than 28,176 feet above the sea, which far exceeds what has hitherto been conjectured.

Hooker was unaware of this fact when he made his own observations of Kangchenjunga. Beyond it, he saw:

> a white mountain mass of stupendous elevation, called by the Nepalis, 'Tsungau' ... it is probably on the west flank of the Arun valley and river ... it is the only mountain of the first class in magnitude between Gosainthan (north east of Kathmandoo) and Kinchinjunga.

He must be referring to Everest, well before its height became common knowledge.

Teething troubles continued with the porters:

> The Bhotan coolies behaved worse than ever; their conduct being in all respects typical of the turbulent, mulish race to which they belong. They had been plundering my provisions as they went along and neither their Sirdar nor the Gorkha soldiers had the slightest authority over them ... I had made up my mind to send back the worst from the more populous banks of Tambur [Tamar], when I was relieved by their making off of their own accord.

The expedition followed a southerly direction to gain the Tamur Valley, the route to the Tibetan frontier. They followed the general direction of the Myong river but kept high above it, travelling along a ridge to avoid the many salients jutting into the valley. As they headed south-west, in the general direction of the cantonments of the Gurkhas, they passed scenery:

> the most beautiful I know of in the lower Himalaya and the Chir Pine is abundant, cresting the hills which are loosely clothed with clumps of oak and other trees, bamboos and bracken.

While passing fields of rice, buckwheat and corn, Hooker enjoyed this tranquil farming country. Always troublesome, however, were:

> the ever-present leeches, mosquitoes, peepsas and ticks which keep the traveller in a constant state of irritation.[1]

The leeches were particularly invasive during the rains; stuffing tobacco leaves down his boots helped to stem their advances.

He and his helpers gradually achieved a happy routine of exercise and rest after the day's exertions. For Hooker, there were tasks to be completed from his somewhat sumptuous tent:

> My tent was made of a blanket thrown over the limb of a tree; to this, others were attached and the whole was supported on a frame like a house. One half was occupied by my bedstead, beneath which was stowed my box of clothes, while my books and writing materials were placed under the table. The barometer hung in the most out-of-the-way corner, and my other instruments all around. A small candle was burning in a glass shade to keep the draught and insects from the light, and I had the comfort of seeing knife, fork and spoon laid on a white napkin as I entered my snug little house, and flung myself on the couch to ruminate on the proceedings of the day and speculate on those of the morrow while waiting for my meal, which usually consisted of stewed meat and rice, with biscuits and tea. My thermometers (wet and dry bulb and minimum) hung under a temporary canopy made of thickly plaited bamboo and leaves close to the tent ...

> After dinner my occupations were to ticket and put away the plants collected during the day, write up journals, plot maps, and take observations until 10 p.m. As soon as I was in bed one of the Nepal soldiers was accustomed to enter, spread his blanket on the ground and sleep there as my guard. In the morning the collectors were sent to change the plant-papers while I explored the neighbourhood, and having taken observations and breakfasted, we were ready to start at 10 a.m.

During the following week, Hooker continued botanising, taking readings and measurements as well as reviewing the geology. He climbed one hill of

---

1   *Peepsa*: a genus of black fly, known to cause river-blindness.

2,835m, where the summit was of broken gneiss rock. This part of the world was for Hooker a pleasant contrast to the gloomy, forest-clad valleys of Sikkim. He observed:

> villages and hamlets appeared everywhere, with plots of golden mustard and purple buckwheat in full flower, yellow rice and maize; green hemp, pulses, radishes and barley and brown millet. Here and there groves of oranges, the borad-leafed banana and sugar cane skirted the bottom of the valleys through which the streams were occasionally seen, rushing in white foam over their rocky beds.

Finally, on 13 November, he reached the Tamur river which the team crossed using a canoe formed from the hollow of a *toon (cedrela)* tree nine metres long. He had previously measured one *toon* tree, two metres above the ground, with a girth of nine metres.

As they gradually gained height above the Tamur Valley, they walked out of the gentle countryside populated by the Limbu and other Kiranti folk and into the mountainous world of the Tibetans. Hooker received a message from Meepo, a Lepcha in a 'smart scarlet jacket'. He had been sent by Campbell to inform Hooker he was now permitted to return to Darjeeling through Sikkim. The *chogyal* (king) had given in to Campbell's persuasions and threats of reducing his pension. This was a considerable relief for Hooker, saving him the long journey south. He now looked forward to experiencing new country all the way to Darjeeling. But first, he had two passes to reach on Nepal's north-east border with Tibet.

On 23 November, the cavalcade had penetrated deep into the Himalaya, to reach Wallanchoon, now part of a village in the Kangchenjunga conservation area.[2] In 1849, it was a very remote, primitive settlement but nevertheless of a surprisingly large size:

> It is elevated 10,385 feet and situated on a fine open part of the Tambur Valley ... very grassy ... the flanks of the mountains covered with luxuriant, dense bushes of rhododendrons, rose, berberis and juniper ... I was almost startled with the sudden change from a gloomy gorge to a broad, flat, populous village of large and good painted wooden houses ornamented with hundreds of long poles and vertical flags, looking like the fleet in some foreign

---

2 Later called Wallungchung Gola and now, on the tourists' map, Olangchunggola.

port; while a swarm of good-natured, intolerably dirty Tibetans were kow-towing to me as I advanced ...The village contained about 100 houses ... built of upright fine strong pine planks, of which the interstices were filled with yak dung.

He describes prayer cylinders, some turned by water, others by hand, *gompas* and banks of *mani* walls. He describes the 'Tibetan tea' they were offered as:

a sort of soup made from brick tea of which a handful of leaves is churned up with salt, butter and soda, then boiled and transferred to the teapot ... which the good woman of the house keeps incessantly replenishing and urging you to drain.

Many yaks were grazing round the village, along with a few ponies, sheep, goats, fowls and pigs, but there was little cultivation other than potatoes, turnips, radishes and some spinach:

Much of the wealth of the people consists in [the yak's] rich milk and curd eaten either fresh or dried or powdered into a kind of meal. The hair is spun into ropes and woven into a covering for their tents ... from the same material are made the gauze shades for the eyes used in crossing snowy passes; the bushy tail forms the well-known 'chowry' or fly flapper of the plains of India; the bones and dung serve as fuel ... their flesh is delicious, much richer and more juicy than common veal; that of the old yak is sliced and dried in the sun forming jerked meat which is eaten raw, the scanty proportion of fat preventing it from becoming very rancid, so that I found it palatable food. The yak ... loves steep places ... [but cannot] bear damp heat, for which reason it will not live in summer below 7,000 feet where liver disease carries it off after a very few years. Lastly, the yak is ridden especially by the fat lamas who find its shaggy coat warm and its paces easy.

Hooker is, like Hodgson, continually making observations about the many different ethnic groups he meets, particularly the Lepcha, whom he greatly admires and the Bhutia, whom he admires less. (The name comes from Bhot, or Bod, the indigenous name for Tibet.) He describes them:

> [the Bhote] inhabit a climate too cold for either the Lepcha or the Nepalis, migrating between 6,000 and 15,000 feet with the seasons, always accompanied by their herds. In all respects of appearance, religion, manners, customs and language they are Tibetans and Lama Booddhists but they pay tax to the Nepal and Sikkim Rajahs, to whom they render immense service by keeping up the trade in salt, wool, musk etc. which could hardly be conducted without their cooperation. They are ... generally poor and very indolent.

After a few days in this village, Hooker and his scaled-down team set off for the first of the passes overlooking Tibet. On 26 November, he left Dharmsala (3,960m) for a sixteen-kilometre walk to the pass; as they ascended, the snow lay deeper until they were ploughing through a one-metre-deep trough which at times became chest deep.[3] Luckily, the path was being used by yaks. They overtook several. Finally, at 3.30 p.m., he and two of his party gained the summit where:

> we were utterly knocked up. Fortunately, I carried my own barometer that indicated 16.206 inches, giving comparative observations with Calcutta, 16,764 feet, and with Dorjiling, 16,748 feet, as the height of the pass. The thermometer stood at 18 degrees and the sun now being hidden behind rocks, the south-east wind was bitterly cold. The plants gathered near the top of the pass were species of *Compositae*, grass and *Arenaria*; the most curious was *Saussurea gossypina*, which forms great plugs of the softest white wool, six inches to a foot high, its flowers and leaves seeming uniformly clothed with the warmest fur that nature can devise.

His comments about the pass are brief, describing a saddle in the middle of which a cairn had been built and festooned in prayer flags. The view beyond into Tibet comprised ridge after ridge of snowy mountains. No doubt with so much scientific work to carry out he had little time or energy at that altitude to express his feelings on achieving his long sought-after goal.

The party then descended to the foot of the pass in about two hours. They continued to a sleeping-place under enormous boulders. Except for an excruciating headache, Hooker felt no other effects. After a supper of tea and biscuits he slept soundly. He had now achieved the first of his

---

3   Dharmsala is not to be confused with Dhar(a)msala in Himachal Pradesh, India, the seat of the Tibetan government in exile.

frontier objectives. The pass at the time was called Wallanthoon, later the Wallang pass, until more recently it has become known as the Tipta La, or Tiptala Bhanjyang, with a height of 5,118m.

On 28 November, Hooker split his party, sending the majority back to Darjeeling via the Tamur with his collection of plants and rocks. He kept nineteen of the most willing personnel to visit another pass in the adjacent valley to the east, the Kang La Chen (also written 'Kanglachen'). After that ascent they would all travel south and east via the Nango, Kambachen and Kanglanano passes to Jongri in Sikkim, on the south side of Kangchenjunga. They set off from Wallungchung with seven days' food, supplemented with a 'scanty supply of very dirty rice at a very high price bought in the local bazaar'.

The parting of the ways came at the junction of the Tamur and the Yangma rivers. Hooker followed the Yangma upstream with his 'chosen ones'. The way was difficult and dangerous, often up ladders and along planks lashed across rock precipices overhanging the turbulent river. Eventually, by 1 December, they had reached the village of Yangma, 'a miserable collection of 200–300 stone huts', with small fields growing radish, barley, wheat, potato and turnip at nearly 4,300m. Sadly for the expedition, the only food available was 'a little thin milk and a few watery potatoes'.

Nearer the pass, Hooker regretted having sent his rifle down as they suddenly came upon twenty-five enormous wild sheep. His description of them, and their horns, suggests that the animals were Himalayan blue sheep, known locally as *bharal*. Hooker named the beasts *Ovis ammon*; there are many sub-species.

The party advanced towards the Tibetan border until mid-afternoon, up snow-covered moraines where Hooker took angles and observations indicating their position to be 16,038 feet (4,890m), still some 1,000 feet (300m) below the Yangma pass. They could make no further progress due to sheer exhaustion and the need to reach the campsite, thirteen kilometres away, before dark. After making various observations for his map and for the theories he had formed on glacial recession, the party headed south down the Yangma Kola until they arrived at a gorge which would lead them to the Nango pass. In crossing it, Hooker made as many geological observations as botanical:

> Granite appeared in large veins in the crumpled gneiss at a great elevation, in its most beautiful and loosely crystalline form – of pearly white prisms of feldspar, glassy quartz, and milk-white flat plates of mica with occasionally large crystals of tourmaline. Garnets were very frequent in the gneiss near the granite veins.

Unfortunately, as the day progressed, the weather turned miserable, with a damp mist turning to drizzling fog which made this first crossing – in soft, deep snow – of the Nango La by a European, 'disagreeable'. The pass's height he calculated at 15,770 feet (4,808m). They descended to camp in darkness; here, they found their first juniper for fuel. On 5 December, after a breakfast of pheasant, snared by Meepo the Lepcha messenger, they descended through forests of Himalayan larch to the Kambachen Valley. Today, this is still the way to the north face of Kangchenjunga and looks as remarkable as Hooker found it in 1848. He estimated the height of Kambachen village at 11,380 feet (3,470m): the village consisted of about a dozen houses on a flat terrace a few metres above the river, surrounded by a few small fields growing radishes, potatoes and barley.

Of all the mountain gorges he had visited:

> this is by far the wildest, grandest and most gloomy; and that man should hibernate here is indeed extraordinary, for there is no route up the valley, and all communication with Lelyp, two marches down the river, is cut off in winter and the houses are buried in snow and drifts fifteen deep are common ... Up the valley the view was cut off by bluff cliffs; whilst down it the scene was most remarkable: enormous black round-backed moraines rose tier above tier from a flat lake-bed ... These had all been deposited at the mouth of a lateral valley opening just below the village and descending from Junnoo, a mountain of 25,312 feet elevation and one of the grandest of the Kinchinjunga group whose top – though only five miles distant in a straight line – rises 13,932 feet above the village. Few facts show more decidedly the extraordinary steepness and depth of the Kambachan Valley near the village, which, though nearly 11,400 feet above the sea, lies between two mountains only eight miles apart, the one 25,312 feet high, the other (Nango) 19,000 feet.

The villagers were most friendly, sharing what little food they had for the winter and supplying a guide to help the team over the Choonjerma pass, (now Mirgin pass, 4,675m) to the Yalloong (Yalung) Valley.

On the following day, they met Tibetan traders, one of whom had snow-blindness. His wife asked Hooker for help. She gave him a present of snuff and carried a little child, stark naked.

I prescribed for the man and gave the mother a bright farthing to hang around the child's neck which delighted the party.

He played a game with his spring measuring tape, which he threw on the ground:

the mother shrieked and ran away, while her little savage howled after her.

As they worked their way towards the Mirgin La they enjoyed fine views of Jannu (7,710m):

Looking north, the conical head of Junnoo was just scattering the mist from its snowy shoulders and, standing forth to view, the most magnificent spectacle I ever beheld. It was quite close to me bearing north-east by east and subtending an angle of 12 degrees 23 minutes, and is much the steepest and most conical of all the peaks of these regions. From whichever side it is viewed it rises 9,000 feet above the general mountain mass of 16,000 feet elevation, towering like a blunt cone, with a short saddle on one side and dips in the steep cliff: it appeared as if uniformly snowed, from its rocks above 20,000 feet (like those of Kinchinjunga) being of white granite, and not contrasting with the snow.

On reaching the Yalung Valley next day, they were told by Tibetan salt traders that the Kang La was impracticable owing to deep snow. They now had to abandon the most direct route into Sikkim and work their way south along the Yalung to the first of those passes which might be open. As they moved downstream, they left behind the giant screen of snow peaks to the north-east: Kabru, Rathong, Kokthang and others which they passed as they marched below the Singalila Ridge. They finally found a pass: on 15 December, they crossed the Singalila Ridge via the Islumbo pass at an elevation of 3,350m and entered Sikkim.

Even after a night without sleep on account of numerous tick bites, Hooker noted the dwarf bamboo, berberis and roses coated in ice. He collected samples of moss before descending to the valley of the Kulhait (Kalet) river, a tributary of the Great Rangeet (Rangit).

After weeks of deprivation, Hooker and his party received a royal welcome at the village of Lingcham:

> The Kajee or Headman had sent out a party with torches to conduct us, and he gave us a most hospitable reception, honoured us with a salute of musketry and brought an abundance of milk, eggs, fowls, plantains and *murwa* beer.

Much news awaited Hooker. More news, however, lay three days away at Yuksom: Dr Campbell had left Darjeeling to meet the *chogyal* of Sikkim at Bhomsong, on the Tista river, which no Europeans had previously visited. Campbell requested Hooker to meet him at Bhomsong and present himself at court to the *chogyal*. Hooker was clearly perplexed for, despite the British having taken back the *chogyal*'s lands which the Gurkhas had occupied, he had:

> for sixteen years steadily rejected every overture for a friendly interview and even refused to allow the Agent of the Governor General to enter their Dominions; it was evident that grave doings were pending ...

The expedition continued for twenty-five kilometres down the Kulhait river valley to cross the Great Rangeet, in order to march over the ridge dividing the Rangeet from the Tista. On the way, Hooker's expedition received generous hospitality from the Lepcha people and from the lamas:

> The villagers had erected a shady bower for me to rest under of leaves and branches, and had fitted up a little bamboo stage on which to sit cross-legged as they do ... after conducting me to this the parties advanced and piled their cumbrous presents on the ground, bowed and retired; they were succeeded by the beer carrier, who plunged a clean drinking-tube to the bottom of the steaming bamboo jug and held it to my mouth then, placing it by my side, he bowed and withdrew. Nothing can be more fascinating than the simple manners of these kind people.

After crossing the Great Rangeet, the pass led past an enormous *mendong*, two hundred metres long, three metres high and twenty-one metres wide. The expedition crossed the divide with the Lapchen at the Lakmo pass. The *kajee*, a local headman, remained at Hooker's side and imparted information, especially on the use of plants, including the fibres of nettles:

some being twisted for bow strings, others as thread for sewing and weaving; while many are eaten raw and in soups.

The pass was measured at 6,800 feet (2,073m). In effect, it was a wide saddle between Mainon (11,000 feet/3,354m) to the north and Tendong (8,663 feet/2,641m) to the south, a fairly major south-east ridge which originates in the Kangchenjunga massif and which finally peters out south of Tendong at the Great Rangeet river.

After a hearty welcome from Dr Campbell, they met the *dewan*, the *chogyal*'s minister who, according to Hooker, was a:

> Tibetan and relative of the Rani (or Rajah's wife); a man unsurpassed for insolence and avarice, whose aim was to monopolise the trade of a country and to enrich himself at its expense.

Despite much subterfuge by the *dewan* to frustrate the meeting, it did take place but without any firm commitments made towards future access between Darjeeling and Sikkim. The *chogyal* was about seventy years old:

> Miserably poor without any retinue, taking no interest in what passes in his own Kingdom, subsisting on the plainest and coarsest of food, passing his time in effectually abstracting his mind from the consideration of earthly things and wrapped in contemplation, the Sikkim Chogyal has arrived at great sanctity.

His followers, and Hooker it seems, venerated the *chogyal*. The interview was soon over, however, with the main benefit being Campbell and Hooker making a good impression on the lamas and villagers who so mistrusted the *dewan*. On Christmas Day, Campbell and Hooker left to climb Mainong (10,630 feet/3,241m) They spent several days on and around the summit, collecting, measuring and recording impressions of the formation of the Himalaya:

> The upper 10,000 feet of Kinchin and the tops of Pundim, Kubra and Junnoo are evidently of granite and are rounded in outline ... The general appearance was as if Kinchin and the whole mass of mountains clustered around it had been upheaved by white granite which still forms the loftiest summits and has raised the black stratifying rocks to 20,000 feet in numerous peaks and ridges ...

The fact of the granite forming the greatest elevation must not be hastily attributed to that igneous rock having burst through the stratified and protruded beyond the latter: it is much more probable that the upheaval of the granite took place at a vast depth and beneath an enormous pressure of stratified rocks and perhaps of the ocean; since which period the elevation of the whole mountain chain and the denudation of the stratified rocks had slowly been proceeding.

Having climbed their mountain and visited the monasteries of Tashiding and Pemayangste, the two friends parted company; Campbell returned to Darjeeling while the indefatigable Hooker headed north to Yuksom, from where he could explore the Rathong Valley and the south side of Kangchenjunga.

He left the picturesque Yuksom village, the last inhabited place before Kangchenjunga, on 7 January 1849, and headed north alongside the Rathong Chu. On 8 January he ascended a very steep mountain called 'Mon Lepcha' which provided extensive views of Sikkim, south to Darjeeling and north, where:

> the eye followed [the Rathong] river to the summit of Kinchinjunga (distant 18 miles), which fronts the beholder as Mont Blanc when seen from the mountains on the opposite side of the valley of Chamounix. To the east are the immense precipices and glaciers of Pundim and on the west those of Kubra, forming great supporters to the stupendous mountain between them.

Hooker took many readings from this marvellous vantage point. (Freshfield, on his circuit of Kangchenjunga in 1899, camped here and christened it 'The Belvedere'.)

Not unexpectedly for midwinter, Hooker was prevented from going higher by copious snowfall while the temperatures plummeted. He gathered his men together and waded downhill through deep snow:

> They took their appointed loads without a murmur and sought protection for their eyes from the glare of the newly fallen snow, some with as much of my crepe veil as I could spare, others with shades of brown paper or of hair from the yaks' tails, whilst a few had spectacle-shades of woven hair, and the Lepcha loosened their pigtails and combed their long hair over their eyes and faces.

After receiving many acts of kindness and spontaneous hospitality, Hooker and his Lepcha attendants found themselves again in Darjeeling, to the delight of their families, who showered Hooker with presents as a mark of their goodwill and their delight at being reunited with their sons.

The second expedition

After spending two months with Brian Hodgson in the Terai, Hooker prepared for his second major expedition, to the north and north-east of Sikkim.

On 3 May 1849, he departed with a party of forty-two, including another guard of five sepoys. The first part of his chosen route required a descent to the Great Rangeet, among sal trees and pines, to 250m above sea level, then over Tendong mountain (8,671 feet/2,644m by Hooker's observations). By 10 May, he had returned to the Tista at Bhomsong. This time, the heavy rain and the obstacles the *dewan* was likely to put in his way provoked gloomy feelings:

> The many difficulties that beset my path all crowded on the imagination when fevered by exertion and depressed by gloomy weather and my spirit involuntarily sank as I counted the many miles and months intervening between me and my home.

This comment appears to show one of Hooker's very few bouts of homesickness.

Several attempts to sabotage the expedition's progress, by dismantling bridges, giving false information, frightening the porters with tales of slavery in Tibet and – the most damaging – denying the expedition access to food, had been made. Very little food relief had arrived from Darjeeling, because of trails being washed away by landslips. Rather than pressing on regardless, Hooker decided to spend ten days in the area of Zemu, exploring the valleys to the north-east of Kangchenjunga. After several valiant efforts to reach the Zemu glaciers he had to give up, mainly on account of monsoon precipitation but also from having little experience of mountaineering. However, he did reach a point less than thirty-five kilometres from Kangchenjunga, at around 4,050m, but was unable to penetrate dense thickets of rhododendrons and steep crags. The final obstacle was a collapsing mass of snow over the Thlonok river (Zemu Chu):

> All my attempts to advance up the Zemu were fruitless and a snow bridge which I hoped to cross to the opposite bank was carried away ... Botany was my only resource, and as vegetation was

advancing rapidly under the influence of southerly winds, I had a rich harvest: although *Compositae, Pedicularis* and a few more of the finer Himalayan plants flower later, June is still the most glorious month for show.

This was Hooker's final foray to the flanks of Kangchenjunga. No one unravelled the secrets of the massif more than he.

He now turned his attention to his natural calling, to stand again on Sikkim's northern border and look across to the great plateau of Tibet. Against all odds, with the *dewan* and his family and agents creating innumerable problems, as well as the very heavy monsoon, on 24 July Hooker stood on the Kongra Lama pass (Kongra La) at 15,745 feet (4,800m), according to his barometer reading:

We were bitterly cold, as the previous rain had wetted us through and a keen wind was blowing up the valley. The continued mist and fog intercepted all views except of the flanks of the great mountains on either hand.

In spite of the height and the cold, Hooker still had the energy to review the botany, where he found:

Isolated patches of vegetation appeared on top of the pass where I gathered forty kinds of plants, most of them being of a tufted habit characteristic of an extreme climate.

During the following three weeks of collecting, he moved to the Lachen Valley in easy stages, as far as its confluence with the Lachung river. On 15 August, having received supplies from Darjeeling, he started up the Lachung to the Tibetan border by yet another pass. En route, he hoped to cross the Donkia pass, to explore the headwaters and sources of the Tista.

This final phase of Hooker's Himalayan odyssey began inauspiciously when his dog, Kinchin, slipped off a bamboo log and was swept away by the Lachung river. Hooker records how much he had become attached to his Tibetan mastiff. He continued north from Nuthung, taking in the hot springs just before Yumesamdong (12,005 feet/3,660m).

On 31 August, the expedition reached Monsandong, which Hooker calculated at 16,362 feet (4,988m). He camped by a yak herder's hut, beyond which hardly any vegetation grew, apart from creeping willow, dwarf rhododendrons and a few prostrate juniper and ephrasia ('eyebright'). On 9 September, he marched north-east from Yumesamdong:

> where stupendous mountains upwards of 21,000 feet high rear themselves on all sides and the desolation and grandeur of the scene are unequalled in my experience.

After passing several shallow lakes and counting thirty or forty different plants on their marshy banks, he ascended more steeply, to arrive at the rocky summit of the Donkia La (5,534m). He recorded the pass's height as 18,466 feet (5,629m) by barometric readings. To the north, the Cholamoo Lake (Tso Lhamo) lay 1,500 feet (450m) below.

Hooker was now well acclimatised after so many weeks of high and continuous exercise. He repeatedly attempted to ascend Kinchinhow, now called Khangchengyao (6,889m), and Donkia from Monay, generally reaching 5,500m to 5,800m. The observations taken on these excursions are sufficiently illustrated by those on the Donkia pass:

> They served chiefly to perfect my map, measure the surrounding peaks and determine the elevation reached by plants; all of which were slow operations, the weather of this month being so bad that I rarely returned dry to my tent: fog and drizzle, if not sleet and snow, coming on every day without exception.

Between Kinchinhow and the Donkia La lies Gurudongmar (5,631m), though it is not mentioned by Hooker. Could it be it was the nearer peak (Gurudongmar) he attempted? It would seem that the peak he called 'Donkia' was, in fact, Pauhunri (7,128m).

On 18 September he visited the Sebu La (5,346m). This pass lies north of Chombu (6,264m) and along the range which divides the Lachen from the Lachung valleys. The view from the summit of the pass:

> commands the whole castellated front of Kinchinjhow, the sweep of the Donkia cliffs to the east ... while to the west, across the grassy Palung dunes, rise Chomiomo, the Thlonok mountains, and Kinchinjunga in the distance.

Hooker's interest in these northern frontier regions extended to more than just their geography: every scrap of information on trade with Tibet and the movement of people and animals to and from Sikkim was recorded.

On 5 October, Dr Campbell and Hooker met at Chungthang, and proceeded once again towards the Tibetan border. By now, Campbell had obtained the authority of the Deputy Governor of Bengal (Lord Dalhousie being absent) to visit Sikkim. They marched up the Tista/Lachen Valley, past Zemu, to follow the Lachen by the same route Hooker had taken in July. They rode to the Kongri La where they were questioned by Tibetan border guards. While Campbell and their companion, the Tchebu Lama, were in conversation with the Tibetans, Hooker galloped off up the Lachen Valley, bent on following the river to its source at Tso (lake) Lhamo. At the lake he rested:

> with the pleasant sound of the waters rippling at my feet. I yielded for a few moments to those emotions of gratified ambition which, being unalloyed by selfish considerations for the future, become springs of happiness during the remainder of one's life.

The next day, they ascended Bhomtso, to the north of the lake. Hooker measured it at 18,590 feet (5,668m). From the summit, it commanded views of all the peaks in the area, including Kangchenjunga, sixty kilometres to the south-west, and Chomolhari to the south-east.

After another ascent of Bhomtso, the expedition left this wild basin for Tumlong, the new capital of Sikkim, via the Donkia La. Unfortunately, after considerable effort, they failed to obtain an audience with the *chogyal*. Hooker continued to pursue his obsession with reaching the Tibetan frontier and on 7 November, with Campbell, climbed to the Cho La, which was completely free of snow at a height, according to Hooker, of 14,925 feet (4,550m). They now stood midway along the ridge which marks Sikkim's eastern boundary and had reached one of many passes crossing what is sometimes known as the Donkia Range, separating Sikkim from the Chumbi Valley. Suddenly, many Tibetan border guards appeared; Campbell seemed anxious to engage in conversation with them and purposefully descended 300 metres on the Chumbi side to their guard posts. Campbell and Hooker were refused permission to descend into the Chumbi Valley. Why they thought they had any chance of doing so without permission is not explained. They climbed back to the pass, where they met several Sikkimese sepoys, sent at the behest of the local governor. They jostled and harassed Campbell; the Tibetans, somewhat ironically, asked them to desist. As the expedition members and the governor's people were settling in for the night Hooker heard Campbell shouting:

'Hooker! Hooker! The savages are murdering me!' I rushed to the door and caught sight of him ... struggling violently; being tall and powerful he had already prostrated a few, but a host of men bore him down and appeared to be trampling on him. At the same moment I was myself seized by eight men, who forced me back into the hut ... here I spent a few moments of agony as I heard my friend's stifled cries grow fainter and fainter ... I struggled but little, and that only at first, for at least five-and-twenty men crowded around and laid their hands upon me, rendering any effort to move useless; they were, however, neither angry nor violent, and signed to me to keep quiet.

Hooker was in fact allowed to go but naturally stayed close to where his friend was incarcerated. It was not until 9 December, nearly one month after Campbell was set upon, that he was freed; the expedition was allowed to proceed to Darjeeling, arriving there on 23 December 1849.

It had been this unauthorised entry into Tibet that had given the *dewan* an excuse to imprison Campbell. Only after threats that the British would send in troops, seize more territory and stop his pension did the *chogyal* cave in and release Campbell, blaming his Chief Minister who then, in turn, did the usual thing in blaming his officials.

(Chogyal Tsugphud Namgyal was to rule Sikkim for sixty-nine years, until his abdication in 1862. He died in 1863, aged seventy-eight, having seen the British annex more Sikkimese territory, including, in 1850, Kalimpong, as a result of this attack on Campbell, the official British government's representative.)

The Sikkimese authorities were naturally worried by a British subject on the Tibetan border. They were also naturally suspicious of the measurements that Hooker had been assiduously making there, obviously conducting a survey intending to produce a map. The map Hooker produced remained for many years the only one of this area (apart from Godwin-Austen's magnificent 1866 chart of the land between Kalimpong and Punakha; it included parts of south-east Sikkim). Hooker's map proved invaluable to future travellers, being of economic and military importance to the British, according to Pema Wangchuk and Mita Zulca in *Khangchendzonga* (2007). It has, consequently, occurred to many commentators on this unfortunate final drama of the Hooker expeditions that it was unfair on the *chogyal*, not only to lose a lot of his territory to British India, but also to lose his pension.

6

Artists, Writers and Photographers

The first artists to sketch and paint the Himalaya were primarily motivated by science. Art, however, was a necessity while photography was still in its infancy. The pencil, pen and palette recorded the geography, flora and lives of local people. The majority of nineteenth-century visitors to the Kangchenjunga area returned with inspiration for others, through literature, paintings and photographs.

Following Sir Joseph Hooker and Hermann Schlagintweit (see below) in the middle of the nineteenth century, Major James Sherwill and his colleagues, now recognised as the first trekkers (see Chapter 7), wrote a vivid account of their explorations. They were followed by serious artists, of whom Edward Lear was the foremost. His magnificent paintings were the first to alert the wider world to the wonders of north-west Sikkim as seen from Darjeeling.

The sketches and paintings of Sir Joseph Hooker (1817-1911) and Walter Fitch (1817-1892)

For nearly fifty years, the descriptions of plants, landscape, villages and villagers which Sir Joseph Hooker (see Chapter 5) brought back from Kangchenjunga were brilliantly interpreted and prepared for publication by Walter Fitch. He was a Glaswegian, just seventeen when he began mounting dried plants. Hooker's father, Sir William, had taught him how to draw plants accurately enough to satisfy his fellow botanists. Fitch worked at Kew when Sir William became Director of the Royal Botanic Gardens.

In Ray Desmond's beautifully produced book *Sir Joseph Dalton Hooker* (1999), three illustrations demonstrate the progress of Hooker's sepia drawing of Kangchenjunga from Singtam to Fitch's watercolour and the final lithograph that appears in *Himalayan Journals*. The result is a very pleasing image of campsite, forest, mist and mountain, superior to any photograph (although there is no mention of Fitch anywhere in the published version).

Other similar sketches, the first from the north-east and north of Sikkim, are similar. The panorama of Tibet and Cholamoo (Tso Lhamo) lakes from the Donkia pass (5,534m) evokes the region's beauty: the snows on the pass and the distant peaks contrast with the brown hills so characteristic of the Tibetan plateau.

Hooker produced the first map of Sikkim, the only one available in such detail for several decades. The map, his fascinating Victorian prose and his insistent sketching were all well received by British colleagues, with the highest praise coming from his friend Charles Darwin. Hooker's *Himalayan Journals* and map were published in 1854, probably just in time to be of use to Hermann Schlagintweit.

Hermann Schlagintweit (1826-1882)

Five years had elapsed from Hooker's visit before the arrival of Hermann Schlagintweit (1826–1882). He was born in Munich, the eldest of five sons. With his brothers Adolf and (later) Robert, he produced early scientific studies of the geography and geology of the Alps. These publications having established their reputations, Alexander von Humboldt, the father of modern geography, recommended them to the East India Company; they were commissioned to scientifically survey the East India Company's lands, and to investigate the Earth's magnetic field. The brothers were also competent artists and mountaineers who became the first Europeans with Alpine climbing experience to visit India. They spent from 1854 to 1857 in the service of the Company and, in order to maximise resources, often split up to visit different regions.

Hermann set out on 5 April 1856 from Darjeeling, intending to traverse the Singalila Ridge northwards, to Kangchenjunga. He had been frustrated by the obstructive attitude of the Sikkimese authorities. On the advice of Dr Campbell, still the Superintendent at Darjeeling, he marched from Darjeeling in two days to climb Tonglu. From there he continued north along the Singalila Ridge as far as Phalut (3,602m), three kilometres south of Singalila Peak (3,636m), where the southern Sikkimese border meets Nepal. His progress was halted here by Nepali border guards. He succeeded, however, in making sketches for paintings of distant peaks, to form a fine record of his visit. John Tucker in *Kangchenjunga* (1955) points out that the main interest in Schlagintweit's visit focused on his paintings and sketches, particularly of the Everest–Makalu group and his suggestion that one of these peaks might be higher than Kangchenjunga.

At the Alpine Club's first summer dinner on 18 June 1858, Hermann and Robert Schlagintweit were invited as the principal guests. Hermann's valuable panorama of Kangchenjunga was purchased by the Alpine Club; it might be construed that some Alpine Club members had, as long ago as 1857, designs on climbing Kangchenjunga. Of more importance, however, is his admirable artistic legacy. In *Kangchenjunga: Imaging a Himalayan Mountain* (2005), Simon Pierse, artist and art critic, tells us that:

> Hermann was also deeply interested in Buddhist culture and was a gifted amateur artist. *Kanchinjinga in Sikhim*, a German oil-print based on a watercolour painted by Hermann Schlagintweit, has certain similarities to drawings of Ruskin, a near contemporary, and like Ruskin's work is imbued with a deep knowledge of the geology of the Himalayan region. Schlagintweit's *Kanchinjinga* rises in the distance above a rocky foreground of glaciated rocks and is a landscape filtered through the eyes – and brain – of a scientist, geologist and climber.

Edward Lear (1812–1888)

Edward Lear arrived in Darjeeling on 17 January 1874. His first sighting of Kangchenjunga produced mixed emotions: he found it not 'a sympathetic mountain ... a rather distracting and repelling whole'. The next day at sunrise it had become 'a glory not to be forgotten', just like his magnificent paintings of the mountain.

Lear had begun making a living from ornithological drawings when only sixteen years old. Unfortunately, his eyesight deteriorated, so he turned to landscape painting. (He is of course also well known for nonsense poetry, publishing collections in 1846 and 1871.)

Lear left Darjeeling after three months, with numerous sketches and photographs to enable him to work on his paintings at home in Italy. Eventually, he finished three now very well-known watercolours of Kangchenjunga (one approximately three metres by two metres). To my untrained eye, he appears to have overcome his initial ambivalent reaction to the mountain.

Lear was invited to India by his friend Thomas Baring, Earl of Northbrook, recently appointed Viceroy of India. He gave Lear a commission to paint whatever he (Lear) chose. He found other patrons, including Lord

Aberdare and Lady Ashburton. Under some pressure on arrival to fulfil these patrons' commissions, he sustained a fall which gave him severe back pain for the remainder of his visit, as well as an asthma attack and serious depression. He did complete the most wonderful paintings of Kangchenjunga, however. No one can view *Kinchinjunga from Darjeeling* (1879), oil on canvas, without 'sublime' coming immediately to mind. As with so many artists, his paintings became enormously valuable after his death: those of Kangchenjunga sell for well over half a million pounds. He was a courageous English eccentric with a huge natural landscape painting talent.

Marianne North (1830–1890)

Marianne North included Darjeeling on one of her many global journeys to places of botanical interest. Private means supported her artistic endeavours, mainly painting plants *in situ* while writing an overview account of the surrounding ecology.

North was born into a distinguished family of ancient ancestry. She spent her early years in Hastings, for which her father was MP. They often travelled round Europe. They frequently visited Chiswick Gardens and Kew, where she was overwhelmed by the sight of so many tropical plants. Specimens were given to her by Sir Joseph Hooker, who became a close friend and mentor. At that time, thanks to Hooker, she began to examine the flora of the tropics.

In 1869, her father died. To assuage her grief through preoccupying herself, she began to plan the first of many lengthy journeys. She had learnt landscape painting in Spain, and thereafter always carried a sketchbook and diary. She remained a watercolourist until an Australian house-guest arrived at their Hastings lodge to give North lessons in oil painting:

> and I have never done anything else since, oil painting being a vice, like dram-drinking, almost impossible to leave off once it gets possession of one.

North exemplified the Victorian pioneering spirit which, with her indomitable physical strength and determination, resulted in well over 800 paintings from her seventeen years of travel.

In 1892, her memoirs and autobiography were published as *Recollections of a Happy Life*, and the abridged version, *A Vision of Eden: the Life and Work of Marianne North*, was published in 1980 in collaboration with

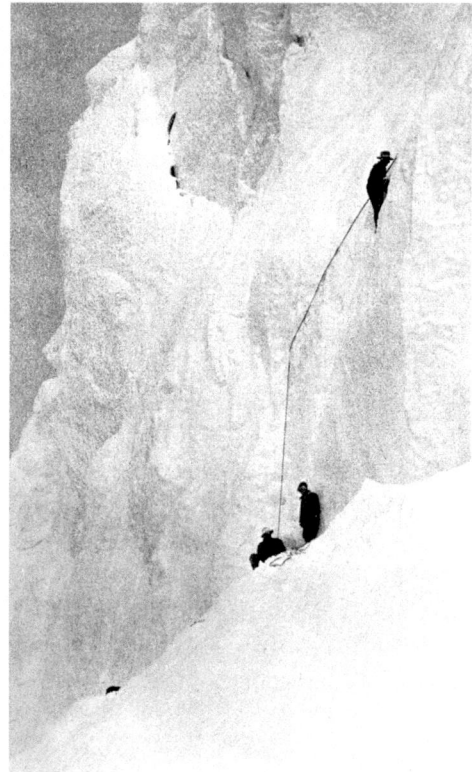

THE 1930 INTERNATIONAL EXPEDITION
Top: Günter and Hettie Dyhrenfurth with team members. © *Frank Smythe.*
Above left: The north-west face from Base Camp. © *Frank Smythe.*
Above right: Climbing the Great Ice Wall. © *Frank Smythe.*

ART AND PHOTOGRAPHY
Top: *Kanchinjinga in Sikhim*, oil print by Storch and Kramer, Berlin, after an original watercolour by Hermann Schlagintweit (1855). © *Alpine Club Collection (NA023P).*
Above: Nikolas Tombazi, from his *Account of a Photographic Expedition to the Southern Glaciers of Kangchenjunga in the Sikkim Himalaya.* © *Alpine Club Photo Library, London.*

North Face, Kangchenjunga (2001), oil on canvas by Julian Cooper. © *Julian Cooper.*

FLORA NEAR KANGCHENJUNGA
Above: Rhododendrons.

Top left and above: Forest.
Top right: Gentians.
Middle right: Primulas.

THE KANGCHENJUNGA MASSIF
Top: The North Face of Jannu.
Above: The Twins and Kangchenjunga from Pangpema.

LOCAL PEOPLE OF THE KANGCHENJUNGA REGION
Top: A family home.
Above: Sherpanis and children from Ghunsa.

THE 1979 EXPEDITION SUPPORT TEAM
Top left: Ang Phurba, sirdar.
Top right: Nima Tenzing, second Sherpa (who was also a porter on the 1955 Kangchenjunga expedition).
Above: Nima Tamang, the cook's boy, at Base Camp.

the Royal Botanic Gardens, Kew. (This highly recommended book is immensely informative.) Her amazing collection of paintings is now displayed at the Marianne North Gallery, Kew.

Included in the Kew exhibition are, of course, her paintings from India and Darjeeling. She gives a favourable account of it:

> the next day took me over the most glorious road, among forest and mountains, to Darjeeling, the finest hill place in the whole world; and I brought my usual luck with me for Kinchinjanga uncovered himself regularly every day for three hours after sunrise during the first week of my stay.

In autumn 1878, North produced the oil painting of Kangchenjunga that can be seen on the cover of Ray Desmond's book on Hooker. The mountain:

> formed the most graceful snow curves, and no painting could give an idea of its size. The best way seemed to me to be to attempt no middle distance, but merely foreground and blue mistiness of mountains over mountain. The foregrounds were mostly lovely: ferns, rattans and trees festooned and covered with creepers, also picturesque villages and huts.

It was typical of her style that the paintings were foreground-framed with plants and trees while the mountains remained distant.

North retired to the Cotswolds, creating a garden populated with plants donated by friends from around the world. Marianne North's sister, Catherine, wrote of her final days in *A Vision of Eden*, when her phenomenal strength was giving out from the punishment of travelling solo, year after year, with so few of the conveniences travellers enjoy today:

> How much her natural stately presence and simple yet dignified manner helped her in facing all sorts and conditions of men in those long distant journeys. She inspired respect wherever she appeared, and good men everywhere were eager to help her.

This fitting testimonial describes a remarkable woman who died too early but left a magnificent public legacy, including half a dozen fine paintings of Kangchenjunga, at Kew.

Mark Twain (1835-1910)

Something similar is perceived in *Following the Equator* (1897) by Mark Twain, who visited Darjeeling during his 1895 tour of the British Empire:

> Everyone on arrival at Darjeeling has much to say and write about Kangchenjunga but not everyone reminds us that there is a view to the south that in its extent is also impressive.
>
> We were well up in the region of the clouds, and from that breezy height we looked down and afar over a wonderful picture – the Plains of India, stretching to the horizon, soft and fair, level as a floor, shimmering with heat, mottled with cloud-shadows, and cloven with shining rivers. Immediately below us, and receding down, down, down, toward the valley, was a shaven confusion of hilltops, with ribbony roads and paths squirming and snaking cream-yellow all over them and about them, every curve and twist sharply distinct.

Twain's acute ability to observe life in the hills is well worth reading. In Darjeeling, he comments on accommodation:

> Up there we found a fairly comfortable hotel, the property of an indiscriminate and incoherent landlord, who looks after nothing, but leaves everything to his army of Indian servants ... I was told by a resident that the summit of Kinchinjunga is often hidden in the clouds, and that sometimes a tourist has waited twenty-two days and then been obliged to go away without a sight of it. And yet went not disappointed; for when he got his hotel bill, he recognised that he was now seeing the highest thing in the Himalayas. But this is probably a lie.

Twain finds a comfortable place at a club and does not go away empty-handed. At his window smoking a pipe, after two hours he is able to see Kangchenjunga:

> I saw the sun drive away the veiling gray and touch up the snow peaks, one after the other, with pale pink splashes and delicate washes of gold and finally, flood the whole mighty convulsion of snow-mountains with a deluge of rich splendours. Kinchinjunga's peak was but fitfully visible, but in the between times it was vividly

clear against the sky – away up there in the blue dome more than 28,000 feet above sea level – the loftiest land I had ever seen, by 12,000 feet or more.

Alfred Williams of Salisbury (1832–1905)

Alfred Williams of Salisbury, the mountaineer-artist, decided to give himself a name-location to ensure he was not confused with another artist of the same name (who also died in 1905). Williams the climber became an Alpine Club member; he was also a founder member of the Swiss Alpine Club and a member of the Scottish Mountaineering Club. Williams concentrated on the mountain without foreground or middle-ground distraction. He often exhibited beside other watercolourists of the day; a memorial exhibition was mounted at the Alpine Club in 1905.

C.B. Phillip (who wrote Williams's obituary for the *Alpine Journal*), considered the high point of Williams's life to be his journey to the Himalaya in 1901, when he was nearly seventy. The last quarter of the nineteenth century had passed without any artist of note taking an interest in Kangchenjunga and the surrounding mountains in Sikkim. Simon Pierse makes an interesting comparison in his *Kangchenjunga* between Williams's large watercolours painted from Darjeeling, including *November Morning at Darjeeling, showing Jannu, Kabru, Kangchenjunga, Pandim*, with Williams's predecessors on the scene:

> The artistic problem of 'huge spaces', that of reconciling the distant – but dominating – backdrop of a Himalayan mountain range with the immediate foreground, and resolving the resulting void of 'nothingness' in between, is one that had been encountered some years earlier by both Edward Lear and Marianne North. North chose to frame Kangchenjunga picturesquely in a vignette of lush jungle shrubbery whilst Lear was at times distracted by what seemed to be the 'impossibility of expressing the whole as a scene'.

Perhaps it is Williams underplaying the middle distance in these paintings, reducing them to waves of colour, allowing the viewer – like the traveller – to concentrate his focus on the distant view, which makes these paintings so modern in their conception. Painting at the very beginning of the twentieth century, Williams was perhaps not so bound by conventions of picturesque composition as his predecessors Lear and North had been.

Williams's large watercolour of Kangchenjunga was painted in the clear light of mid-morning after the mountain forms had been drawn out in some considerable detail. Pencil marks are visible on the surface of the paper where they were painted around in Chinese white or body colour, creating the forms of the mountains and allowing the off-white surface of the paper to show through.

Nicholas Roerich (1874-1947)

I met Russian friends in a Moscow flat, to discuss a climb in the Fann mountains of Tajikistan; there, I first saw a Roerich print of the Himalaya. Such was my interest that the flat's owner produced a set of postcard-sized prints depicting a dozen or more of his paintings. I was made to take them home with me. The sublime use of colour alone ensured a second look, but they also gave out an unearthly quality despite being bold in shape and form. They were obviously the work of a master who knew the significance of the abode of the gods and of Kangchenjunga in particular, 'the treasure house of the spirit'.

Roerich's work evoked such memories when first seeing the mountain from Darjeeling, early one spring morning. At dawn, after a period of heavy rain to clear the air, with the light from the east prominently defining each of the massif's peaks, and with the orange light contrasting with the dark blue mountainsides still in shade, that view is breathtaking, as are Nicholas Roerich's interpretations of it.

Nikolai Konstantinovich Rerikh was born in St Petersburg on 9 October 1874. He spent his early years in the family home on the Nikolaevsky embankment with his older sister and two younger brothers. The name Roerich means 'rich in glory' and the estate name, Isvara, is a Sanskrit word meaning 'divine spirit or lord'. (A previous owner of the estate had travelled in India.) 'Nikolai' means 'one who overcomes': altogether, an auspicious start!

The family spent happy summer months at their estate's manor house where a gravure of a snow-covered mountain lit up by the setting sun remained forever in Roerich's memory. He eventually discovered that the peak was the sacred mountain Kangchenjunga, a mountain he was to depict in several magnificent works after seeing it from viewpoints between Bengal and Darjeeling during the 1920s.

In 1918, he and his family left Russia for Finland during the Bolshevik Revolution. In 1919, he moved to Britain, where he hoped to obtain entry permits for India. In 1923, he set sail for India, arriving in Darjeeling towards the end of December. He and his wife took some time to locate a suitable home, only to find it had been occupied by the Dalai Lama when he fled

Lhasa to escape the Chinese in 1910: Nicholas and Elena could hardly be blamed for seeing a causal force which propelled them to India, to paint Kangchenjunga and the mountains of Central Asia. The Roerichs ended their days in the Kullu village of Naggar, the site of the Roerich information centre and museum.

Roerich seems to have identified most closely with the people who live near the mountain and who intuitively grasp its spiritual significance. He was, by 1923, steeped in oriental religions and teachings and, through Buddhism, he had become acquainted with 'Shambhala', a mythical, hidden, holy place, thought to be somewhere in the highlands of Central Asia.

Many artists and critics believe that Kangchenjunga inspired Roerich's greatest works. From the wonderful viewpoints around Darjeeling, he painted a series of twelve canvasses, called *His Country*, inspired by the holy men who had come to build an ashram in Sikkim. This is mentioned in *Altai-Himalaya* (1929), although Roerich does not identify the ashram. This book reads well with its companion volume, *Shambhala* (1930), both books having been published shortly after his return from the heart of Asia. Roerich applied that creative energy to some of the finest mountain landscape paintings imaginable, none more so than his renderings of Kangchenjunga.

Howard Somervell (1890–1975)

In the spring of 1928, Howard Somervell returned to Darjeeling and Sikkim for a mountain holiday, with a friend, Bill Allsop. He had previously travelled through this part of the Himalaya as a member of the 1922 and 1924 Everest expeditions, on which Somervell had distinguished himself by climbing without oxygen to about 8,500m.

With ten of Somervell's former Everest porters, they marched north from Darjeeling eventually to reach the Guicha La with the intention of climbing Pandim, 'the nearest major Himalayan peak to civilisation which had never been climbed'. They realised it was much too formidable a mountain for their two-man team. They did discover an unmarked valley and glacier before heading to the Kang La, but by then Allsop, having contracted dysentery, was too ill for further mountaineering.

Somervell wrote that their trip achieved nothing but 'provided us with many hours of deep enjoyment and a few moments of paradise'. He also produced several sketches and a watercolour entitled *Southern Aspect of Kangchenjunga*, depicting the valleys south of the Guicha La seen from the slopes opposite Pandim. During this excursion, he also painted the classic view from the Singalila Ridge of Gaurisankar, Makalu and Lhotse, before returning to work as a medical missionary in the south of India.

Theodore Howard Somervell was not only a mountaineer and doctor but an accomplished musician and artist. His father, William Somervell, was an early influence on his painting, encouraging his son to sketch at every opportunity, to the point where his family realised they were nurturing a compulsive artist.

Somervell indicates that Nicholas Roerich influenced his painting. He visited Roerich's house in Kullu for a few days in 1945, as he records in the *Himalayan Journal* (1946):

> Professor Nicholas Roerich, the finest mountain painter now alive, has got the knack of getting all the dignity of the hills on to his canvas, by deliberately letting only the really significant lines of the peaks and glaciers and rocky foregrounds be reproduced by his brush, very often in a simple curve or line.

For Somervell, Roerich was 'the greatest mountain painter alive'. Both artists produced memorable Cubist works, in depicting surfaces, light and shade through use of various geometrical figures.

Somervell's paintings are only partly documented as his output was prolific: he gave away many of his works to friends and family. He was not a commercial artist like Edward Lear, which, as David Seddon observes:

> allowed his style to develop much more freely ... Probably no other artist applied Cubism to the high mountains in such a consistent and authoritative way as Somervell ... Most would regard his paintings of the great Himalayan peaks and Tibet as unique and they are an important part of the heritage of the Alpine Club as well as the history of mountain art.

Of the hundreds – maybe over a thousand – of paintings which Somervell produced, Seddon's research enabled him to account for 540 paintings of which 201 are of the Himalaya or Tibet. The Alpine Club, in January 1923, mounted a major exhibition, including fifty-seven works by Somervell, of which twelve were of his travels and climbs in Sikkim while on the 1922 Everest expedition. They include Fluted Peak, Siniolchu and Jonsong Peak, all part of the Kangchenjunga massif.

Somervell painted Kangchenjunga mainly from forays into Sikkim and north-east Nepal on his return from the two Everest expeditions. Altogether, Seddon has identified four watercolours and three oils of Kangchenjunga. Many others remain untraced:

> I have traced four watercolours of Kangchenjunga. *Kangchenjunga from below Darjeeling* (1925) is of particular interest in that it bears the inscription 'EFN from THS. 1925'. Edward Norton remained a close friend of Somervell after the Everest expeditions, and Somervell may have given this painting to Norton to mark the publication of *The Fight for Everest: 1924*. Another watercolour, *Southern Aspect of Kanchenjunga* (1928), depicts the valley to the south of the Guicha La from the slopes opposite Pandim and was exhibited at the Alpine Club in 1929. *Kangchenjunga from Tiger Hill* is dated 1947 and another, *Sunrise over Kangchenjunga from Darjeeling*, is undated. All are in private collections; the last was the Alpine Club's Christmas card in 2004. Two more watercolours, *Domed Peak, Kangchenjunga* and *Forked Peak, Kangchenjunga*, as well as *Pandim, Sikkim* and *Kabru, Sikkim*, were exhibited at the Club in 1936.[1]

Simon Pierse in *Kangchenjunga: Imaging a Himalayan Mountain* tells us that Somervell was a member of the Lake Artists' Society, where William Heaton Cooper (1903–1995) was influential. Heaton Cooper's son, Julian, trekked through Nepal to study and paint facets of the mountain 'that bring us close to the unapproachable'.

Somervell helped widen our understanding of Kangchenjunga: his contribution to the mountain and humanity goes far beyond his artistic compulsion. He was President of the Alpine Club from 1962 to 1965. I was asked to present a lecture there, 'Climbs in the Lands of Islam', in 1967. This was my first visit to the Club; I was somewhat surprised that all those on the first few rows were soon asleep, some snoring loudly, as I droned on, until I mentioned the *North-north-east ridge* route of Toubkal, 'which was climbed by the redoubtable Bentley Beetham, who some of you may have known'. At that point an elderly, stocky gentleman awoke, stood up, thanked me for my comments and declared, 'I knew him well.' He continued for a few minutes to say just how well, then sat down to resume his sleep. He was, I discovered later, Howard Somervell. It was a moment not to be

---

1   *Alpine Journal*, 2005.

forgotten, 'one of the hardiest mountaineers of all time', as Noel Odell wrote in Somervell's obituary in the *Alpine Journal* of 1976, adding that 'It will be long before the world sees again the like of this many-sided and accomplished man.'

Julian Cooper (1947-)

Another Lakeland artist visited the Kangchenjunga region more recently, almost a hundred years after Somervell first travelled there. Julian Cooper arrived with fresh views on what was and was not important in landscape. Not for him the classic view from Darjeeling: he wanted to go to the heart of the mountain massif. In late autumn 1999, he walked to Pangpema, the yak pasture at just under 5,200m above the Kangchenjunga glacier which also serves as the base camp for climbing on the north-western flanks of Kangchenjunga and its outlying peaks. He arrived with his long-time rock-climbing partner, Alan Howard, his assistant in carrying canvasses and easel, paints and cameras into position, so Cooper might paint the surroundings using the *plein-air* method, to best capture the effects of atmosphere and light on the topography. Here, in the shadow of Kangchenjunga, what he could see of the mountains' flanks through binoculars was what interested him most, the heart of his new approach to mountain painting.

Julian Cooper's grandfather, Alfred Heaton Cooper (1863–1929), was a post-Impressionist landscape painter of note. His son, William, produced watercolours which he used to illustrate his Lake District books as well as the Fell and Rock Club rock-climbing guides which, though containing only the classic routes, are still sought after for the Heaton Cooper illustrations.

As Andrew Lambirth remarked in his essay *Julian Cooper: Reading the Rock*, in the Wordsworth Trust brochure *Mind Has Mountains* to support the 2001 Wordsworth Trust exhibition at Grasmere, 'Tradition can be as much a burden as a benefit.' By the time he had progressed through Lancaster Art School, Goldsmiths College of Art, then taken advantage of travel scholarships for painting across Europe and steeped himself in esoteric literature, he had returned to Cumbria to read the rock, to climb and paint steep places as no one had before.

Cooper's major breakthrough came after he had travelled through Nepal to the north-west of Kangchenjunga, having been inspired to produce oil paintings of Chang Himal, Wedge Peak, Jannu north face, Sobithongie, Kabur, Drohmo and Kangchenjunga north face. This latter painting is topographically accurate, clearly showing the 'Croissant' where, from a final bivouac, we made the third ascent of the mountain. Most of the painting covers a vast area of the mountain's hazardous northerly aspect: the snow-fields and ice cliffs which avalanche relentlessly, day and night.

In an interview with Grevel Lindop, the poet, art critic and former Professor of Romantic Studies at Manchester University, Cooper discusses his Kangchenjunga expedition paintings. These are notable for absence of summit and sky and instead focus on selected facets of the mountainsides which are scrutinised minutely before producing images as accurate yet abstract as they are original. Cooper's images of that small portion of the steep mountainside fix in the mind; his name is now synonymous with a unique interpretation of mountain landscape. The paintings serve up a feast of the features which make up the mountain vision – rock, ice, snow, air and light – most skilfully.

Stephen Venables, when President of the Alpine Club, conveyed a perceptive impression of Cooper's use of paint.

> Julian Cooper's monumental canvases focus ever closer on the sculptural form and texture of landscape; in his Kangchenjunga series, gorgeously layered impasto echoes the fantastic encrustations of Himalayan snow fluttering, the paint oozing and dripping to suggest the constant gravitational flux of glacial ice.[2]

Cooper's explorations of Kangchenjunga's mountainsides represent a quest towards knowing what it is that draws him and indeed all of us towards such mountains. 'The mountains have this immense presence,' says Cooper. 'There's an energy about them that doesn't come from somewhere outside. It's inherent in them.'

'Painting is a bit like alchemy, I suppose,' Cooper muses. 'It's one of the last human activities where people mess about with material stuff and hope for something to happen – something that isn't material at all.'

Photography on and around Kangchenjunga

It was natural that British photographic enthusiasts should travel to India. After a slow start during the 1840s, because of the difficulty in obtaining chemicals and the vagaries of climate affecting the exposures, photographic studios had been established in all the main cities by the next decade.

Prominent among the photographic entrepreneurs who tried to record the mysterious east was Samuel Bourne, who had become a very successful photographer while working in a Nottingham bank. He set up three studios,

---

2   Foreword to *The Artists of the Alpine Club* (2006) by Peter Mallalieu, Keeper of the Pictures.

in Simla, Bombay and Calcutta, from where he gained a reputation for high quality landscape photographs, after processing the result of plates made on three Himalayan expeditions between 1863 and 1866.

Theodore Julius Hoffmann and P.A. Johnston opened as commercial photographers in Calcutta in 1882, with a branch in Darjeeling which became one of the largest photographic operations in India. Many other photographic studios opened, to cater for the demands of visitors from the plains of India and tourists from abroad. They supplied not only black and white prints of the mountain villages and local people but also, in some cases, paintings. During the final decades of the nineteenth century, picture postcards became popular and provided the most significant source of income for these studios.

Photography soon played a big part in government surveys, from archaeological surveys to recording ethnicity. This resulted in an eight-volume work, *The People of India* (1868–1875), with many contributions from the indefatigable portrait photographer of local people, Benjamin Simpson (1831–1923). Surgeon General for the Government of India from 1853 until 1890, Simpson was also a member of the Bengal Photographic Society. He produced eighty photographs of 'Racial Types of Northern India', for which he was awarded a gold medal at the London International Exhibition in 1862.

Major Laurence Waddell (1854–1938)

Waddell travelled through Sikkim and Tibet, as Chief Medical Officer of Sikkim, between 1886 and 1896. To illustrate his book *Among the Himalayas* (1899), Johnston and Hoffmann printed images on glass plates which he had portered across the Himalaya with great difficulty between 1888 and 1896.

Waddell, during his many journeys north from Darjeeling into largely uncharted country south and east of Kangchenjunga, recorded on his glass plates the local people, their settlements and the topography. His book contains early first-hand accounts by a European of those born and raised near Kangchenjunga, including the Lepcha, the original ethnic inhabitants. Their way of life makes fascinating reading, although it was not at first an easy task to take photographs of these people who had strong reservations about the box pointed in their direction. They believed:

> It took away their souls with their pictures and so put them in the power of the owner of their photograph to cast his spells over them and similarly a photograph of the scenery they alleged blighted the landscape.

Waddell offered presents and, with his no-doubt disarming Glasgow humour, usually managed to persuade his subjects to overcome their scruples.

John Claude White (1853–1918)

Claude White, the first Political Officer to be appointed to Sikkim, complemented the journeys made by Waddell. White, too, was as interested in the local people as the landscape. He helped to open up routes to the distant peaks, as it was his responsibility to construct roads and bridges. On a journey during June 1891 to the Zemu Valley on the east side of Kangchenjunga, he was accompanied by Theodore Hoffmann, whom he commissioned to take photographs. Several of the images appear in *Among the Himalayas*, including peaks such as Siniolchu seen from the Zemu glacier and a clear view to the west of Kangchenjunga towering above every other mountain.

White maintained an abiding interest in photography from his early days in government service, particularly during his assignment in 1883 to the British Residency in Kathmandu, from where he photographed the unique architecture and the now famous monuments of the Kathmandu Valley. Later, in Sikkim, using his glass plate camera, he recorded for the first time landscapes and portraits of the region's people. John Falconer, curator of photography at the British Library's Oriental and India Office collections, rated Claude White as:

> probably one of the last, and certainly among the most impressive products of a tradition of quasi-amateur photographers which had flourished among administrators and military personnel in India since the 1850s.

The Himalaya around the Kangchenjunga massif excited Claude White the most. He wrote:

> There is something very exhilarating in these high altitudes, the tremendous expanse of snow around gives a feeling of freedom not experienced at lower elevations, while there is always a fascination at arriving at a summit of a mountain, particularly where the unknown is on the other side.

Kurt Meyer's exhibition of much of Claude White's photographic record in 2005 was titled *In the Shadow of the Himalayas*.

Vittorio Sella (1859-1943)

Sella took some of the finest images of mountains ever. Those he took in 1899 of the Kangchenjunga massif revealed these mountains in all their magnificence as no other photographer had before. He achieved this when camera equipment weighed a great deal, restricting the range of activity of most photographers. Sella was fortunate in many respects, as his birthplace and upbringing favoured him developing a love of the mountains and his dedication to recording them, as if he was predestined to do so.

He was born and died in the foothills of the Alps at the north Italian town of Biella. Similar to many of the artists who came to paint Kangchenjunga, Sella was influenced by a family interest in the art of photography. His parents were well off, a circumstance that enabled his father to pursue his photographic interest and publish in 1856 the first handbook in Italian on photography. Naturally, Vittorio developed an early awareness of and appreciation for photography as well as for composition, since he received an education in painting.

A family tradition of mountaineering helped. In 1863, his uncle, Quintino Sella, an accomplished mountaineer, brought together other committed practitioners to form the Italian Alpine Club. Vittorio Sella was therefore an obvious choice when Freshfield was organising his tour of Kangchenjunga. His letter of invitation was more casual than persuasive but tempted his man to join his team:

> Dear Signor Sella,
> Do not trouble yourself overmuch by my suggestion. I am an uncertain person balancing possibilities – if I can see my way to leave England for 6 months I should be more disposed to do so. Could I hope to get you to bring your equipment and experience, I am very lazy and inefficient in making elaborate preparations. Still it is just worth our while to turn it over in our minds as an idea ...
> I should like to see those great peaks and to go round Kangchenjunga. I have thought so for twenty years and now it is getting, perhaps it has got, too late!
>
> Yours truly,
> DWF Douglas Freshfield, March 1899[3]

---

3 *Summit: Vittorio Sella – Pioneer mountaineering photographer 1879 to 1909* (1999).

Sella, with help from his brother, Erminio, and their assistant Erminio Botta, calmly went about the task of exposing 196 glass plates of 20 x 25 centimetres, using his cumbersome Dallmeyer and Ross wood-bodied cameras set on a much heavier tripod.

With his companions and porters, he carried cameras and ancillary equipment round Kangchenjunga; many of his photographs complemented Freshfield's narrative of the seven-week circuit, *Round Kangchenjunga* (1903). The photograph entitled *Below the Jonsong La – Nepalese side* appears, at 6,000m, to have been the highest taken by that date. It clearly shows the upper north ridge and north face of Kangchenjunga with cloud blowing off Jannu on the far right. Many more of the images taken on this expedition constituted the first record of the topography of north-east Nepal; certainly, no European had stood on the Jonsong La before 1899.

Similarly, Sella's images of the local people, particularly of the shy Lepcha, are also wonderfully composed historical records of a now all but extinct ethnic group.

His photographs were widely praised. The famous American landscape photographer Ansel Adams declared they inspired:

> a definitely religious awe … in Sella's photographs there is no faked grandeur; rather there is understatement, caution, and truthful purpose.

The majority of Sella's mountain views occur under a natural sky, not blackened by filters but remaining grey, to give the spectator the feeling of being there in the space and light permeating the air and mountainsides. The only exception is his portrait of Siniolchu summit, taken by telephoto from the Zemu glacier just after a massive snowstorm. To capture the symmetry of the mountain and the splendid incrustation of its precipices in a fretwork of snow and ice with the Himalayan sun burning down and reflected on to his glass plates, the darkened sky was inevitable. Freshfield concludes his appreciation of Sella's photograph and the mountain: 'Siniolchu is, for the climber, the ideal snow mountain; the throne where power dwells apart in its tranquillity'.

'Remote, serene and inaccessible': more so than any other photographer, Vittorio Sella was able to bring that mood home from the Eastern Himalaya, to inspire others to visit and to help mountaineers to plan their climbs on and around Kangchenjunga.

PART 3
Climbs and Attempts

7
The Pioneers

During the nineteenth century, societies were founded to encourage leisure interests such as mountaineering: the Alpine Club was founded in 1857. Thanks to the stimulus of the Great Trigonometrical Survey and the expansion of the British empire in India, moreover, professional and amateur mountaineering interests extended beyond the Alps to the Himalaya.

Because Kangchenjunga stands prominent for all to see, it has exerted a powerful influence, particularly on those who live in sight of it. In fact, Kangchenjunga has become the focal point of a nation: no wonder it has inspired climbers to appreciate its grandeur and ethereal nature, seemingly so accessible to all.

Major James Lind Sherwill (fl. c.1860)

Major Sherwill, with three companions, Dr B. Simpson of the Bengal army, Captain E. MacPherson of the 93rd Highlanders, and W. Kemble, a Bengal civil servant, left Darjeeling following a week of fair weather on 2 November 1861. (Sherwill realised that November is an excellent time to trek in the Himalaya, despite Sikkim's high precipitation.) Their aim was to ascend:

> any one of the perpetually snow-clad spurs of the great Kanch-unjingah group of mountains and [examine] the glaciers of this hitherto unexplored portion of the great Himalayan range.

Following the Treaty of Tumlong (1861), they were the first mountain trekkers to obtain legal access to Sikkim.

Reaching the perpetual snows via the crest of the Singalila Ridge had been discouraged since 1852, when Captain Walter Sherwill had been prevented from making progress along the ridge by a 'deep and precipitous valley'. This group, however, hoped to attain their goal via the Rothang river after consulting Dr Hooker's recently published map of Sikkim.

They rode north out of Darjeeling to the boundary between British and independent Sikkim, then proceeded on foot through forests of 'stately Saul trees' and 'groves of giant bamboos'.[1] Towards the end of a very long first day:

> one of our party became quite knocked up by the long and fatiguing walk but after despatching the best part of a tin of marmalade, was sufficiently recovered to proceed and mount the remainder of the steep acclivity.

The next day, the party continued north through mainly cultivated land of millet, rice and buckwheat, using a new road established the previous year by sappers of the British Indian army during its temporary occupation of this part of Sikkim.

The cavalcade of Lepcha, Limbu and Bhutia helpers advanced north, supplementing their rations by purchasing food from friendly villagers and monasteries. They passed the ruins of the once very extensive Pemayangtse monastery, which had burnt down the previous year. The weather remained fine. Dr Simpson took photographs of the snow peak to the north from the fine vantage point of the monastery, situated at 2,130m, before descending north-west to the Rothang river:

> a wild, foaming and boiling current, dashing over large blocks of gneiss rock ... The temperature of the water was 48 degrees.

A march uphill brought the party to Eksum (1,780m), a village now grown into a town called Yuksom, the 'meeting place of three learned monks'. (In 1641, Lama Lutsum Chembo, Lama Sempa Chembo and Lama Rinzin Chembo travelled from Kham in Tibet to propagate Buddhism in Sikkim. This trinity elected Phuntsog Namgyal as the first king of Sikkim, giving him the title *chogyal*, which could be interpreted as 'he who rules with righteousness'.)

Sir Joseph Hooker (see Chapter 5) wrote at some length about Yuksom: on 11 January 1849, he had walked for three hours towards Kangchenjunga until poor weather obliged him to retreat to Darjeeling. Major Sherwill and

---

1   Saul (now usually 'sal') tree: the Indian dammar, a hardwood with multiple uses.

his friends, however, continued in good weather, past Hooker's furthest point, before deciding to ascend Gubroo (Kabru, 7,412m). The attempt was begun from the south-east side, at Kabru Lake. They found the going:

> trying from its steepness, and the great elevation causing shortage of breath, nausea and violent headaches. We reached about 16,500 feet when I found it impossible to proceed any further in consequence of an oppression in the head and a feeling like that of sea-sickness.

Having left all their heavy baggage cached with porters, they then set off on a four-day excursion to go where no European had penetrated before. They made their first camp close to the formidable peak of Pundim (Pandim, 6,691m). That evening, having made the baggage 'coolies' comfortable in their tents, Sherwill had time to reflect that:

> the grandeur of the surrounding snow-clad mountains ... surpasses anything I have elsewhere witnessed in the Himalayas.

The following day, they passed *mendongs* and crossed mounds of glacier moraine to reach several small lakes. They then (sensibly) returned to their camp for the night and the next day set off for the head of the valley, where the ice and moraine became ever steeper. Near the summit of a pass:

> Kemble was nearly precipitated to the bottom by his foot giving way and only saved by rapidly digging his Alpine stick into the snow, which pulled him up.

On reaching the northernmost point of this valley, well above the Onglaktang glacier, they found themselves at 5,640m, looking down to the Talung glacier and beyond to the hugely impressive south flank of Kangchenjunga, now only twelve kilometres distant. Despite the intense cold, the doctor succeeded in taking photographs from the pass, including one of the north face of Pandim, later to be known as the Guicha (or Goecha) La (4,940m).

The party retreated to Darjeeling in orderly fashion, having extended the exploration of the Kangchenjunga massif by going thirteen kilometres further than Sir Joseph Hooker. They returned wiser men about acclimatisation and avoiding the worst effects of acute mountain sickness. Their trek had lasted twenty-one days; they had enjoyed perfect weather, apart from occasional light afternoon mist. A full account of this journey can be read in the *Journal of the Asiatic Society of Bengal* (1862, pp.457–479).

Sherwill's party was the first to go trekking, purely for pleasure, in the Eastern Himalaya. Just as he stood on the shoulders of his predecessors, others benefited from reading his account.

Elizabeth Mazuchelli (1832-1914)

The first woman to reach the summit of Kangchenjunga, Britain's Ginette Harrison, did not do so until 1998. Interestingly, during the early years of exploring Kangchenjunga, a 'Lady Pioneer' played a prominent part in the second expedition to make 'a tour in the interior of the Himalayas', out of curiosity rather than for any scientific reason.

India Office library records give Elizabeth Sarah Mazuchelli's date of birth as 1832 and that of her husband, Francis Mazuchelli, as 1821. The Langford history group in their *Every House Tells a Story* project discovered that Francis was born in Milan, emigrated to the USA where he was ordained into the Catholic church and in 1849 was awarded the degree of Doctor of Divinity. In 1853, he left the Church of Rome then married Elizabeth Sarah Harris in Geneva before returning to England. In 1857, the couple set off for India, where Francis became an Assistant Chaplain in Calcutta and went on to serve in India for twenty years. The highlight of that stay, at least for his wife, spanned the two years of his appointment to Darjeeling, with its more temperate climate, and which included their memorable two months crossing the 'Indian Alps'. To avoid vulgarity, in her account of their journey, *The Indian Alps and How We Crossed Them* (1876), the author refers to herself as a 'Lady Pioneer'.

Elizabeth Mazuchelli explored, in 1876, the western side of the Kangchenjunga massif during a two-month excursion from Darjeeling, on foot, horseback and sometimes carried by *dandy-wallahs* with a retinue of bearers, guides and cooks which her husband Francis and his friend, 'C' (not otherwise named), had organised.[2] The first reference to this visit, as far as can be ascertained, is to be found in *Round Kangchenjunga* (1903), where Douglas Freshfield in his bibliography describes the book as a:

> Trip along the Singalela ridge in winter, probably as far as the Semo La – a trivial and topographically obscure narrative, well-illustrated.

---

2   *Dandy-wallahs*: men who carry hammocks or litters, often for Western tourists.

Douglas Side wrote for the 1955 *Alpine Journal* an excellent and comprehensive article, 'Towards Kangchenjunga', where he mentions Mazuchelli's 1876 journey. He is somewhat perplexed, however, as he writes:

> The story published anonymously in 1876 as *The Indian Alps and how we crossed them*, is difficult to follow either with an old, or with a recent map. The journey was made in winter, along and across the Singalila Ridge. On the map published with the book, the route is shown as terminating above the snowline on the southern slopes of 'Junnoo'. The descent to Tseram was made from or near Kang La.

(The *Alpine Journal* editor, Tom Blakeney, found that the book is listed in the British Library catalogue under 'Nina Elizabeth Mazuchelli'.)

Louis Baume in his *Sivalaya* (1978), still a useful source-book for historians of the Himalaya, refers to the journey. His comment is typically succinct:

> 1876 Mrs. Elizabeth Sarah Mazuchelli (who was generally called Nina) made a journey to the southern slopes of Jannu having come via the Kang La and Tseram after crossing the Singalila Ridge.

The only other reference appears in Kev Reynolds's excellent Cicerone guidebook, *Kangchenjunga: A Trekker's Guide* (1999), where, in his chronology of events for 1876, the entry states, 'Crossing into Nepal by the Kang La, Mrs Elizabeth Sarah Mazuchelli approached the south side of Jannu.'

Mazuchelli's book, still available from antiquarian bookshops, disproves Freshfield's accusations of triviality and obscurity. It is well worth reading: it is very perceptive about the Sikkimese, life in the jungle and the spellbinding effects of Kangchenjunga bursting upon the senses after storm clouds have parted.

The Victorians' fascination for the supernatural appears when Mazuchelli experiences spectral fantasies of 'unearthly vapour' around Kangchenjunga, through which she 'almost entered some new world'. Consequently, it has been said that this book is half-serious, half-comic, yet an altogether absorbing story. Sections of it are highly amusing, written throughout in elegant Victorian prose and, as Freshfield noted, the illustrations are excellent – not just the watercolours but also the pencil sketches of the lady on horseback at the edge of a cliff while looking over her shoulder with some concern at the huge drop below; of her languishing in a *dandy*; of the people met on the trail; life in camp in the jungle; on the mountain – all wonderfully complementing the text.

The text, it must be conceded, is of its time, for example in describing 'my faithful bearers', whom, however, she respected, and they her. Mazuchelli empathised with their problems and rendered first aid whenever accidents or illness befell her retinue. But it is the depth of understanding, the everyday detail of travelling through rugged country that is refreshing to read, since it comes from a woman's perspective. It is so different from Freshfield's writing, or that of Frank Smythe (who fails to mention Mazuchelli's journey in *The Kangchenjunga Adventure*, 1930).

Mazuchelli's book is extremely perceptive in describing the locality's different ethnic groups, beginning in Darjeeling where she records that:

> Its native population numbers upwards of 20,000, consisting of various tribes, Bhootias, Lepchas, Limboos and Goorkhas; the three former having originally migrated from some province in Thibet. They are, for the most part, an inoffensive and peace-loving people, particularly the Lepchas, a nomad race, natives of Sikkim, who possess many virtues ...

She further asserts:

> Very different in each respect are the gentle Lepchas, who are truthful and honest to a singular degree, those who have had transactions with them declaring that seldom if ever have they known them commit a theft or tell a lie. Their complexion is fair and ruddy, but of that yellowish tinge observable in all the Mongolian races, and, like the Chinese, they are oblique-eyed and flat-faced, giving one the idea that they must have been accidentally sat upon when they were babies, and that they have never got the better of it since.

On trek, she echoes Hooker's comparison:

> The Bhootias, as the stronger party, generally have it all their own way, claiming pre-eminence as their right whilst the peaceful Lepcha, who are not prone to wax valiant in fight, yield to them naturally as they would do to everybody. Not so the Nepaulese, who do so but with ill grace.

Having written intimately of her feelings for the locals she does not, however, give much away about herself and her early life.

We do know that she was very relieved to escape from a miserable existence as a Company wife in the oppressive heat of Calcutta. It was, she declared, 'a liberation itself to be in Darjeeling' and more so when her determination 'to have a near view of them [the mountains]' was realised. Which is no doubt why she talked her husband into their two-month journey into the Himalaya, to the valleys descending from the south side of Jannu.

Douglas Freshfield (1845–1934)

Douglas Freshfield organised and led the most important investigation of Kangchenjunga after the first reconnaissance by Joseph Hooker. In doing so, Freshfield came fully to appreciate just what a ground-breaking visit Hooker accomplished. Freshfield consequently dedicated his account, *Round Kangchenjunga*, to the great botanist.

Indirectly, Hooker lured Freshfield to the north-west of Sikkim. On the map attached to Hooker's *Himalayan Journals* (1854), Freshfield noted:

> A broad blank separates travellers' routes on the north-west and north-east of Kangchenjunga. Across the empty space is printed the following stimulating sentence: 'This country is said to present a very elevated, rugged tract of lofty mountains, sparingly snowed, uninhabitable by man or domestic animals.'

Freshfield was consumed by an ambition to fill in that blank on the map by making a circuit of Kangchenjunga.

The person Freshfield failed fully to acknowledge – and certainly not in the wholehearted manner in which he complimented Hooker – is Rinzin Namgyal, who was the first to make a circuit of Kangchenjunga, from October 1884 to 31 January 1885, some fifteen years before Freshfield. This has not gone unnoticed. Pema Wangchuk and Mita Zulca, in *Khangchendzonga: Sacred Summit* (2007), comment with some justification that Rinzin's journey was indispensable to the success of the first European circuit of the mountain. Wangchuk and Zulca's view is:

> It is obvious that Freshfield ... was finding it difficult to balance his sense of Western superiority vis à vis the natives and the obvious brilliance of what Rinzing had accomplished a decade before him

supplied with barely a fraction of the facilities that Freshfield's entourage was lavished with.

Douglas Freshfield, by the end of the nineteenth century, had become one of the most widely active of exploratory mountaineers. He was born in London. His father was a notable lawyer. His mother, Jane Crawford, was the daughter of William Crawford, MP for the City of London, who had reaped huge rewards from employment with the East India Company. His mother was also an author, having published *Alpine Byways* and *A Summer Tour in the Grisons*.

Round Kangchenjunga was immediately recognised as a very important contribution to Himalayan literature. It is a masterpiece of information on the opening of Kangchenjunga to the mountaineering fraternity:

> The Alpine Club recognised Freshfield's contribution to Alpine and Greater Ranges mountaineering, and to the Club, particularly the *Alpine Journal*, by inviting him to become President from 1893–95. Freshfield was above all an exploratory mountaineer, preferring snow and ice and having no natural aptitude or liking for rock. He much preferred to climb on mountains where no one had gone before. Tom Longstaff acknowledged Freshfield's 'mountain sense' which Freshfield put to positive effect in clarifying unmapped topography.[3]

Ironically, after writing the book of which he was most proud, *Round Kangchenjunga*, the publisher's warehouse caught fire; very few copies of Freshfield's *magnum opus* survived.

In 1898, Freshfield began his preparations to circumambulate Kangchenjunga by gathering his team and letting them know what kind of expedition he expected. In a letter of invitation to the most famous mountain photographer of the era, Vittorio Sella, he wrote that he intended the expedition to be:

> A party of pleasure so constituted to produce results that might afford entertainment and even instruction to a larger circle of friends of mountains.

---

3   *A Century of Mountaineering* (1957), Arnold Lunn.

Sella and his brother, Erminio, accepted Freshfield's offer, as did Edmund Garwood, professor of geology and Alpine climber, who had plane-table surveying experience from exploring on Spitzbergen (Svalbard). A Valtournanche guide, Angelo Maquignaz, and a photographic assistant from Piedmont, Erminio Botta, of whom Freshfield does not say more in the book but was, in fact, Sella's usual assistant, also joined the expedition. In India, they were joined by Rinzin Namgyal ('R.N.'), who had made the clockwise circuit of Kangchenjunga. He had been in British government employment since 1878; in fact, on the journey around Kangchenjunga he was to receive 100 rupees per month for revealing the lie of the land in this last unexplored corner of Sikkim and north-east Nepal. The expedition was greatly helped by having at its disposal Mr Dover, the Superintendent of Roads, as well as the Deputy Commissioner of Darjeeling and Captain Le Mesurier, the Political Resident in Sikkim, who was deputising for Claude White. The Deputy Resident absented himself from office work in Gangtok to go to Lachen for ten days to assist with communications, 'and to prevent the coolies from bolting'. They were also accompanied by half a dozen Sikkim Gurkha police.

The Surveyor General, Gore, made available to Freshfield a copy of a sketch map of the north and north-western flanks of Kangchenjunga which had incorporated information supplied by Rinzin. It was apparent that detail was lacking when it came to depicting glaciation. So, for Freshfield:

> To get round Kangchenjunga was not therefore the only object I set before me. I hoped to be able to obtain, what the Indian Survey had been too fully employed elsewhere to be able to give geographers, a fairly accurate general delineation of the main glacial features of the group and some material for comparing them with those of the Alps and the Caucasus.

There were other objectives of the expedition such as recording faithfully 'the landscape that may be serviceable to Alpine climbers'. This achievement helped when making the third (1979) ascent of the mountain (see Chapter 16). Freshfield had planned to ascend to various high points and peaks for the purpose of reconnaissance, but 'rope and ice axe played but a very subordinate part in our journey. This was our misfortune rather than our fault.'

The six European members of the expedition assembled in Marseilles, from where they set sail on 10 August 1899. They arrived in Darjeeling on

1 September after a twenty-hour journey by train from Calcutta. The long-awaited renowned view of Kangchenjunga never materialised from behind the veil of mist and cloud during their four days at the hill station. They would have to wait until their return seven weeks later.

(The tremendous storm that devastated the tea gardens around Darjeeling in late September deposited three feet of snow on the mountain and almost brought the expedition to a premature conclusion; for Freshfield and the members, however, it enhanced their determination to plough a trail over the passes.)

The expedition set off from Darjeeling at midday on 5 September. The party descended into subtropical country where, according to Freshfield, 'each tree has to fight for its share of sunshine'.

They emerged from the virgin forest to continue on horseback to Kalimpong and Gangtok, then on to the former capital of Sikkim, Tumlong. At Gangtok they were guests of Captain and Mrs Le Mesurier, where Freshfield and his friends:

> chaffered with pedlars straight from Lhasa for Tibetan curios; we helped to entertain the Raja and his pretty little Tibetan wife at afternoon tea.[4] She arrived in a palanquin borne by servants in striped kilts and scarlet tunics, wearing conical straw hats decorated with peacocks' feathers.

It had rained during most of the journey, affording no views of the snowy heights:

> What we saw was all beneath us, dripping forest slopes, the profound green abysses of the Teesta Valley, swathed in vast lurid sheets of moving vapours, distant crags and spurs ... between the clouds ... Out of the depths of the forest came companies of mild-eyed Lepchas, slender, timid figures, recalling irresistibly the poet's Lotus-eaters.[5] By birth denizens of the woods, they are by habit and necessity naturalists, they know the ways, and haunts, and voices of all its beasts and birds, the properties of all its plants. May our British officials be able and willing to protect and preserve them from what I fear must otherwise be their inevitable fate, to be superseded by the more sturdy Tibetan and the more energetic Nepalese.

---

4   Chaffered: haggled.
5   'The Lotos-Eaters' (*sic*), a poem by Alfred, Lord Tennyson, first published in 1832.

Freshfield, although feeling inadequate about conveying the beauty and variety of the forest, wrote:

> I must confess to having felt on this ride the same sort of delight a child feels on its first visit to the pantomime. I waited breathlessly for what would come next, and what came was always beyond my expectation. In the open, clusters of thatched cottages rose among green terraced rice fields, or nestled between orange trees, plantains, and feathery clumps of gigantic bamboos. But it was on entering the forest that the true enchantment began. We rode through an endless colonnade of tall trunks – oaks, chestnuts, magnolias, their stems and branches fringed with parasitic ferns and festooned with orchids and creepers. Tree ferns raised their crowns over the carpet of greenery and blossom that covered every inch of ground. Hydrangeas were common, and a yellow convolvulus romped over everything. Down each ravine sparkled a full torrent, making the flowers and ferns nod as it rushed past them. Magnificent butterflies, some black and blue, others gorgeous flashes of colour, fluttered across the sunlight.

The expedition spent two weeks passing through this jungly country in the monsoon, crossing turbulent rivers in full spate to reach Green Lake by the Zemu glacier. It had taken five days for 'our pioneers to cut a track through the rhododendron jungle' from Zemu Bridge and to make camp by Green Lake.

A further aim of the expedition was to climb a peak of around 6,000m, to afford a view of the Nepal Gap on the north ridge of Kangchenjunga, in the hope that it might reveal a way to the western side of the massif. Freshfield and Maquignaz reconnoitred the Upper Zemu glacier. Maquignaz remained asleep at a rest stop while Freshfield continued alone across the ice at about 5,200m, to below the buttress at the start of the east spur of Kangchenjunga. It was oppressively hot; blue sky disappeared behind an ominous yellow haze. Freshfield, collecting Maquignaz en route, hurried back to Green Lake camp where snow began to fall, then hardly stopped for forty hours, casting a mantle more than a metre deep. (The same storm had already caused many problems along its route from the Bay of Bengal and particularly over Darjeeling. It was discovered later that almost seventy centimetres of rain had fallen in thirty-eight hours, sweeping away houses, outbuildings, tea gardens and a native bazaar. Some 500 people in and around Darjeeling died in that storm, which was reported worldwide.)

All thoughts of climbing high were abandoned in favour of continuing the circuit of the massif. The leader could be satisfied that he had reached a point just below the buttress rising out of the Upper Zemu glacier that led to the east spur of the main peak and, in doing so, had gained a position only eight kilometres from the summit. Freshfield had thus surveyed the route for the second ascent of the mountain, although at the time he was not optimistic about its viability. He could also see from his high point, when looking directly south, that there were easy snow slopes leading up to the north side of the Zemu Gap at c.5,850m.

The morning of 26 September broke frosty with blue skies, conditions which remained with only occasional interruptions for the remainder of their tour. In this welcome clear weather, Sella and his assistant went to work with their cameras. Their main objective was Siniolchu (6,888m), with Kangchenjunga set in the background. Vittorio Sella enjoyed a wonderful opportunity during this time to photograph all the mountains in their new coat of snow, with none so magnificent as Siniolchu, despite it being 1,700m lower than Kangchenjunga:

> Siniolchum [sic], owing to its symmetry and proportions, and also to the splendid incrustation of its precipices in a fretwork of snow and ice, has impressed the few Europeans who have as yet approached it as the most superb triumph of mountain architecture.

The cavalcade was now unlikely to receive any support or supplies for several weeks, so it was imperative to move across the Thangchung La (4,980m) before exhausting their food and fuel. They stopped for the night at a camping ground used by local nomads, only 300 metres above the Zemu glacier. Garwood had succumbed to heatstroke, having been over-dressed in his Spitzbergen attire. The following day, they reached the pass, where the cairns were adorned with prayer flags.

Freshfield went ahead on the next day, believing he was following a trail under the snow marked by 'stone men' (route-marking cairns) but, much to the amusement of his companions, they turned out to be giant rhubarb (*Rheum mobile*) plants. They managed to camp by the Tumrachan Chu on flat ground free of snow.

Here, Freshfield reviewed the porters' situation. Some of his observations will be familiar to explorers of more recent times. It seems he had suffered:

the charge of callousness or inhumanity from certain stay-at-home critics in respect of his treatment of porters. No one is safe from such libels, and I think it prudent therefore to take this opportunity to state explicitly the conditions on which our porters were serving us. The men were all volunteers, and they were all paid at a higher rate and better fed than is usual in Sikkim. We provided for them tents, snow-boots and snow-spectacles. Their loads were carefully weighed, and all possible pains taken to distribute them fairly, an endeavour they daily did their best to frustrate by every imaginable ruse. They were never during the whole journey called on to do more than what would be a short half-day's journey for an Alpine peasant. On one occasion they proved their powers by doing voluntarily a long ten-hours' day, two ordinary marches, on a very rough and fatiguing path, in order to reach a village and enjoy a night's carouse before, as they hoped, we would catch them up. This proof of their capacity they judiciously reserved till the end of our travels. Those who were feeble were relieved of their loads as the journey proceeded, those who were ill or frostbitten were doctored by Signor E. Sella. Their weaknesses and their wiles were manifold.

He later added:

We had spared no pains to see that they started well provided for such an adventure. Besides snow-boots and snow-spectacles they had 'portantinas' or frames of the most approved description to carry their packs on. But some had lost their boots, and some their spectacles, and most had used their carrying frames for firewood.

They now had to cross the The La (*sic*). Freshfield, having overtaken the 'coolies', sat behind a rock to rest and watch as:

The procession advanced, chatting cheerfully so long as they were unconscious of my presence. Suddenly they caught sight of me; their cheerfulness disappeared as if by magic. Some fell on all fours, some fell flat on the ground under their burdens, all groaned piteously. Then I laughed outright, and several of them joined in, like children found out in a game.

According to Kenneth Mason in *Abode of Snow* (1955), it was General Bruce's impression that Freshfield did not altogether appreciate the contribution made to the success of the expedition by his porters. Perhaps, writes Mason:

> [Freshfield] was handicapped, as so many visitors from Europe are handicapped, by not being able to communicate directly with them or to understand fully their point of view. Confidence and training alone were lacking in pre-Everest days.

Freshfield was ambivalent in his comments about the role of Rinzin on the expedition, implying that his laid-back attitude in 1899 was evident fifteen years before on his first circuit of Kangchenjunga. Freshfield refers to Rinzin's 'curious behaviour when travelling with us' and goes on to say:

> I am disposed to attribute some of the defects of his survey [1884] to his obvious predilection for sitting in a snug tent and filling in neat, but somewhat subjective, detail. Scrambling with his plane-table among rough ice and moraines, and out-of-door sketching were far less to his taste.

Pema Wangchuk in *Khangchendzonga: Sacred Summit* (2007) points out, in defence of his countryman, that he:

> was already middle aged by that time and even though Freshfield himself was fifty-four years old when he undertook the expedition, he was driven by science and the glory of becoming the first Westerner to have circumambulated Kangchendzonga. Rinzing was already in semi-retirement, a well-to-do landlord who had already achieved what Freshfield had set out to do. There was no real motivation for him and he was perhaps only accompanying the expedition as a favour to his acquaintances in British offices in Darjeeling. That said, Freshfield would never have completed the circuit had he not taken Rinzing along as a guide and this comes across in his narrative, *Round Kangchenjunga*.

In fact, Freshfield writes in Appendix C of his book, in the introduction to 'The Narrative of the Pundits', about 'their bold spirit of adventure and their remarkable powers of topographical and general observation.'

From the The La (*sic*) they descended easily to the Lhonak Chu, a flat-bottomed valley, to walk west, upstream, to where it becomes the Lungma Chu. Erminio Sella, Rinzin and Freshfield climbed to the Chorten Nyima La (5,819m), situated on the watershed, where they could look into Tibet proper. They were aware that Tibet had laid claim to this region; the expedition was continually anticipating being accosted by Tibetan troops. Freshfield was sympathetic to the Tibetan perspective: 'The physical boundary of the watershed does not correspond to the racial limits.' Historically, Tibetans for generations had brought their animals to graze in this northern region of Sikkim. Freshfield, when putting on his geographer's and patriot's hats, favoured the boundary being pushed back to the main Himalayan divide and enclosing all of Sikkim under British influence.

The party climbed south-west after Rinzin became uncertain about the route he had taken in 1884. He remained, for a time, confused, thinking he had come into Sikkim from Nepal via Ramthang and the Chabuk La and then down the Lhonak glacier. Freshfield eventually worked out that this was probably where Sarat Chandra Das and Lama Ugyen Gyatso had crossed in 1879 before crossing the Chorten Nyima La into Tibet on their visit to the Tashi Lhunpo monastery.

Freshfield's team had now arrived at the crux stage of their journey in terms of height, remoteness and unusually deep snow. They had to camp twice on snow, with the higher camp at 6,100m, just short of the crest of a col below the Jonsong La. On 6 October, they crossed this pass and began the equally arduous descent of the Jonsong glacier, where several 'coolies' became frostbitten. Only two days later, Freshfield was informed of a porter who, being 'weary of life', asked his comrades to leave him in the snow; they had left him with a blanket and some biscuits and had bidden him farewell.

The whole party reached Pangpema (5,141m; on some maps, 'Pangperma'), the site of many an expedition's base camp in the future, where:

> Our camp was established on a green shelf between the mountain and the trunk glacier, which was hemmed in on three sides by mighty snow-peaks ... I have adopted from Rinsing's sketch-map of this district the name Pangperma for the spot we had reached. Reference to our own map will show its importance, both for topographical and picturesque purposes. It is situated at a point where the glacial drainage of the greater part of the north-western face of Kangchenjunga unites with that of the chain extending to the north as far as the Jonsong Peak. Four distinct ice-streams, each

made up of many tributaries, here unite to form a trunk glacier of the first magnitude ... His [the climber's] chief difficulty would be on the lower steps of the cirque; here I believe he should search to the left towards the saddle that connects Kangchenjunga and the Twins. There are rocks for a bivouac on the high plateau, and the final climb would be practicable, if the difficulties of altitude do not supervene ... The whole face of the mountain might be imagined to have been constructed by the Demon of Kangchenjunga for the express purpose of defence against human assault, so skilfully is each comparatively weak spot raked by the ice and rock batteries. I failed at the time to trace any route on which skill could avert this danger, and, with Mr Mummery's fate before our eyes, this approach to Kangchenjunga cannot be recommended, even to the boldest climbers, until such a route has been discovered.[6] I am not prepared to say that this may not be done.

Vittorio Sella found it a pleasure to take panoramas of what is one of the most amazing mountain settings anywhere, with five glaciers converging to feed into the Kangchenjunga glacier:

The Himalayan giants are, with a difference, greater Alps; a glacier is always a glacier; but the scale was far larger, and the impression left on the mind one of stupendous vastness ... It is no wonder that the Nepali yak-herders who penetrate to this spot should regard it as the special haunt of the Spirits of the Mountain, a place where Gods and Saints dwell in great numbers.

The party left Pangpema to walk on yak pasture and moraine, now some 150m above the shrinking ice, alongside the Kangchenjunga glacier, all the way to the rough huts at Lhonak (4,816m). Freshfield expressed his views on this name:

The proper name of the place seems to be Ramthang. Rinsing calls it Lhonak on his map; a name which I reject, as it must tend to confusion with the district of that name.

---

6   Mummery's fate: Albert F. Mummery, a prominent Himalayan climber, had disappeared quite high on Nanga Parbat in 1895.

On 10 October, the expedition was comfortably camped on the sparse pasture at 4,000m below Kangbachen (7,903m). From there, easy walks provided excellent vantage points to observe and photograph the towering peak of Jannu, just one of many that Freshfield reconnoitred. He and his team of cartographer and guide had conducted a thorough reconnaissance, not only of the north-western approaches to the summit of Kangchenjunga but also of the western flanks of the peaks to the north: 'Nepal', 'Tent' and 'Pyramid'.

Freshfield acknowledges that fifty years previously Hooker was the first European to visit the Kangbachen Valley, but makes no mention in his book that twenty-five years later Elizabeth Mazuchelli travelled through this area.

They continued down the now verdant yak pastures where:

> A rustic deputation met us ... The farmers wore a broad hearty smile; the women, forewarned by some woodcutters of the strange arrivals had put on all their jewellery, their amber, coral and turquoise ornaments; what was more to the purpose they had brought milk and potatoes. They were the first inhabitants we had met since leaving Lachen twenty-five days before.

They received a welcome at Ghunsa but were being reminded they were now trespassers in Nepal, when 'an embodiment of the law' arrived from a lower village while they were resting. He had attempted to prevent the expedition from obtaining provisions from the people of Ghunsa but the 'present of medicines for his cold in the shape of whisky got over his scruples'. The inhabitants of Ghunsa had not seen another European in the village since Joseph Hooker, but they remained unperturbed by this sudden encounter. It is always possible that Nina Mazuchelli and her entourage had been there, although it is very difficult to know from her narrative exactly where her party went. Her images of Jannu strongly suggest she travelled in this vicinity. They left this 'flourishing and most picturesque village' to continue south by the route the Ghunsa folk sometimes take after collecting butterflies to sell to the Europeans in Darjeeling.

They first crossed the Sinon (Senon/Sinion) La (4,661m), then the Mirgin La (4,676m; referred to as Choonjerma Pass by Hooker). A final unnamed pass of 4,721m followed, before descending to the Yalung Valley. The party crossed the Yalung Chu by a bridge upstream of Tseram, to gain the valley leading up to the Kang La ($c.4,973$m) on 15 October. They had now crossed into Sikkim, to spend five days on or around Jongri (Dzongri, 3,960m). Cloud and snowfall had already prevented them from climbing

Kang Peak (5,579m) or from exploring the glaciers descending from the peaks lying between Jannu and Kabru.

Fine weather returned after they reached Jongri, where they enjoyed fresh food, having been resupplied by Mr Earl, the Deputy Commissioner of Darjeeling, accompanied by a police escort. In clear but much colder weather, they completed a three-day excursion to and from the familiar Guicha (or Goecha) La (4,940m) and the summit of Kabur (4,826m). They considered also climbing one of the summits of Kabru (7,318m) but, having found snow conditions difficult up to the Guicha La, the idea was abandoned.

The party had now carried out a brief reconnaissance of the south-east segment of the Kangchenjunga massif which, as far as they could see, offered no easy way to the summit. From Kabur they were able to view the south face, which they found less threatening; it was not written off. This expedition failed to explore very closely the east flanks of the mountain south of the Zemu Gap, although it was noted that reaching the Zemu Gap from the south side 'looks exceedingly steep'.

The final night of their grand tour was enlivened at Jongri when they lit a huge pre-arranged bonfire that could be seen as far away as Darjeeling. Everyone now knew that Freshfield and his party had succeeded in their enterprise. Soon after, word spread to Calcutta: their success was celebrated with an official gun salute on the orders of the Governor of Bengal.

Towards the end of the expedition, Freshfield speculated about how the mountain might be climbed in the future:

> The foot of the eastern ridge can easily be reached from the Zemu Glacier but the climb of 9,000 feet along it will stop ordinary mortals. The southern cliffs are in appearance hopeless. The obvious key to the upper part of the mountain is the northern ridge, and the best way to reach it seems to be from the Kangchen glacier. It is possible that a very careful and close inspection may reveal a way of getting past the lower icefalls without incurring too great a danger from avalanches ... Higher up, the long steep *névé* stream is broken by two short rockfalls which appear surmountable. Above there is a 'Grand Plateau', with crags near it, which would have to furnish sleeping quarters. Some 1,200 feet more would still have to be climbed. But the last ridge looks practicable. There is a tower on it, perhaps 100 feet below the top. When this is passed the climbers will know they have conquered.

Freshfield makes interesting comments on 'the vexed question of the rarity of the air'. He expresses scepticism about the scientific studies and predictions being carried out at the time, mainly because, in his opinion, everyone behaves differently at altitude. He derides the 'consequent worthlessness of much dogmatic and so-called scientific writing on this obscure and intricate physiological problem'.

He does make a sensible observation: that it is wise fully to acclimatise up to around 5,250m and then, from his experience, most people can climb to 6,000m and above without too much distress:

> On the whole my conclusion is the same as Sir J. Hooker, that, for a considerable proportion of mankind, there will prove to be no impossibility in the attainment of heights up to 22,000 feet, so long as attention is paid to diet, the attitude when in repose, and if the march is regulated and not hurried. I must add that there seems to me to be no sufficient reason for thinking that climbers may not attain 29,000 feet.

Freshfield's friend and brilliant cartographer Edmund Garwood sums up his leader's qualities:

> Freshfield was extremely wiry and a notably fast and untiring walker. He was extraordinarily light for his height, and I remember that on our return to Darjeeling he weighed well under ten stone. He was averse to unnecessary burdens and I cannot recall any occasion on which he was carrying a rucksack – indeed he wore no special mountaineering outfit and his usual attire was the grey cutaway tailcoat and trousers that he habitually wore in England ... I cannot refrain from expressing my admiration to the way in which he planned the expedition after previously making himself acquainted with all that was known about the district. Of his good temper and cheerfulness, under occasional trying and difficult circumstances, and of his unfailing thoughtfulness and consideration for his companions, I cannot speak too highly.

Rinzin had been sent ahead to Darjeeling two days before the main party left Jongri and the mountain. At every *gompa*, large and small, they were greeted with great fanfare and hospitality, none more so than at Yuksom, the first seat of power in Sikkim, and then at the most famous monastery of

'Pamionchi', known as Pemayangtse to the Sikkimese and so well described by Joseph Hooker. Rinzin, having alerted the monks to the imminent arrival of the travellers from Kangchenjunga, a Devil Dance was arranged:

> The occasion was propitious, as a party of young Lamas was staying at the monastery previous to going up to Akukthang to offer a week's service to the God of Kangchenjunga.

After seven weeks away, during which time they had 'ascended and descended some 75,000 feet, or fourteen vertical miles up, and as many down', they rode into Darjeeling where Kangchenjunga, at the end of October, had become visible:

> Every morning the great mountain, calm and radiant, greeted at sunrise his worshippers and, if its summit was withdrawn for a few hours behind the midday vapours, it was sure to reappear after sunset when the veil of twilight had already spread over the lower hills, glowing like a beacon-light set on the verge of another and less material world.

8

The Kabru Controversy

The first mountaineers with any Alpine experience to climb in the Himalaya were Adolf and Robert Schlagintweit, who in August 1855 reached 6,780m on Ibi Gamin in the Kamet group.

The Hungarian Maurice de Déchy, accompanied by the Swiss (Meiringen) guide Andreas Maurer, arrived in Darjeeling in 1879 for no reason other than adventure. De Déchy climbed Phalut (3,602m) on the Singalila Ridge, the same place that Hermann Schlagintweit had reached in April 1855, but could go no further as he had contracted malaria; the expedition was abandoned. It might have been this Hungarian alpinist who was the first to set out to climb in the Himalaya simply for sporting reasons but, in 1883, it was an Englishman who became the first person to succeed in climbing a Himalayan peak for its own (sporting) sake.

William Woodman Graham (1859–c.1932)

In March 1883, the recently qualified barrister William Woodman Graham, with many respectable Alpine routes under his belt, left Darjeeling for Kangchenjunga with guide Joseph Imboden of St Niklaus, 'more for sport and adventure than for the advancement of scientific knowledge', as he famously stated in his Address to the Royal Geographical Society.

In July 1882, the Sella brothers and three guides had fixed ropes to iron stanchions up the Dent du Géant. Graham, with two Chamonix guides, swarmed up these fixed ropes, then reached the higher north-east peak of the Géant. This was considered unsporting enough for the Alpine Club to reject, by a wide margin, Graham's subsequent application for membership, a process which coloured some opinions in the ensuing Kabru controversy. Glyn Hughes, the Alpine Club's Honorary Archivist, revealed this in his article with Willy Blaser for the 2009 *Alpine Journal*, entitled 'Kabru 1883: A Reassessment', an invaluable source for what follows.

Graham and his guide based themselves at the summer pasture of Dzongri, south of Kabru. From the hut, they crossed into forbidden Nepal via the Kang La (5,054m), where they claimed to have climbed a peak of 6,100m to the north of the pass, probably Kokthang (6,147m). They returned to Sikkim where they crossed the Guicha La but decided that conditions were wrong for climbing above the Talung glacier. They therefore retreated – encouraged by a porter having accidentally burnt Graham's boots! Imboden contracted a fever and decided to return home; he was also homesick. He was replaced by another guide, Ulrich Kaufmann, and joined by Emil Boss, the joint proprietor of the Bear Hotel in Grindelwald and an officer in the Swiss army.

Graham and Boss were correct in their criticism of the Indian Survey maps of the Garhwal (where they had been climbing before returning to Kabru). They suggested the Swiss army should help with the training of Indian Survey officers, being 'the producers of the finest cartographical work in the world'. This criticism and remedy did not go down well with the hard-working officers of the Survey of India who, being conscious of their élite status, from then on were unlikely to support Graham in any future debate about the veracity of his claimed ascents.

Graham and Boss returned to Calcutta, Darjeeling and Sikkim at the end of the monsoon. Despite poor weather, they conducted a reconnaissance of the west (Nepal) side of Kabru, which they found discouraging. On 6 September, they walked up the valley leading to Kabru's south side, to hunt for mountain sheep (*Ovis ammon*) which they failed to locate. They decided that the south side of Kabru was more difficult than the west on account of a 'mass of broken glacier'.

On 19 September 1883, the party made an abortive attempt on Jubonu Peak and, on 23 September, they crossed the Guicha La, hoping to find a reasonable route up Pandim (6,691m) from the north. They found that too difficult so gave up, reversing their journey over the pass. They then succeeded in climbing Jubonu (sometimes now called Jhopunnu, 6,524m), reaching the summit on 1 October after Kaufmann led the way at a great pace. 'He is,' wrote Graham, 'generally admitted to be one of the fastest stepcutters living and this day he fairly surpassed himself.' They considered this climb their most difficult so far.

Their main objective for this Himalayan excursion was Kabru (7,412m), the prominent outlier peak to the south of Kangchenjunga. On 6 November, they began their approach to the only possible route remaining, the east flank. They climbed 250m up steep moraine to make camp during a snowstorm. The next day, after finding themselves on a subsidiary buttress, they wasted time reversing and attempting a different route to the north, where they found a ledge just wide enough to accommodate their Whymper tents. Their porters sat outside all night (luckily it was quite mild) and everyone retreated to safety the next morning. The intrepid trio started again, roped, at 4.30 a.m., by crossing a couloir of loose avalanche-prone snow, followed by two hours of snow-cutting and then, 'nearly one thousand feet of delightful rock work forming a perfect staircase'. By 10 a.m., they had reached a point only 450m below Kabru's east summit. The last slope was:

> pure ice at an angle from forty-five degrees to sixty degrees ... owing to the recent heavy snow and the subsequent cold it was coated three or four inches deep with frozen snow and up this we cut notches for the feet.

Thanks to that snow cover, Kaufmann was able to lead them up to the lower summit of Kabru by 12.15 p.m. After admiring the views, including Everest, 120 kilometres away to the north-west, they hurriedly left for the actual summit via a short arete that:

> rose in about 300 feet of the steepest ice I have seen. We went at it, and after an hour and a half we reached our goal ... The absolute summit was little more than a pillar of ice and rose, at most, thirty or forty feet above us still. But, independently of extreme danger in attempting it we had no time. A bottle was left at our highest point and we descended.

The descent was achieved with great difficulty yet, curiously, on the way down they fixed a large Bhotia flag to a smooth slab. They reached camp by moonlight at 10 p.m., 'having been nineteen and a half hours on foot'. This was a fantastic achievement. Apart from it being the highest point ever reached, it was also one of the more difficult peaks to have been climbed by then in the Himalaya.

This ascent, however, was to initiate endless debate. None of the three climbers entered further into the controversy nor added anything to their original report and route description. The most vociferous in arguing against Graham's claims were mountaineers with a hidden agenda who would gain in some way from proving he was mistaken, particularly those who had also climbed high but not quite as high as Graham and his friends.

On 9 June 1884, Graham read a paper describing his visit to the Himalaya to the Royal Geographical Society. It was later published in the *Proceedings of the Royal Geographical Society* and in the *Alpine Journal*. On that night at the Royal Geographical Society, Graham's personal comment on reaching 'an unusual height' on Kabru included:

> Neither in this nor in any other ascent did we feel any inconvenience in breathing other than the ordinary panting inseparable from any great muscular exertion.[1]

The audience was said to have been enthralled by Graham's laconic account of his climbs. Everyone, that is, except Sir Joseph Hooker, according to his biographer, Ray Desmond, who questions whether it was jealousy or resentment that brought Sir Joseph to his feet, obviously perplexed by Graham's assertion that he had hardly suffered at all from the altitude. He declared he had never known:

> what it was to go a few miles outside his tent without feeling great pressure or to walk up to 18,000 feet without a feeling of having a pound of lead on each kneecap, two pounds in the pit of his stomach and a hoop of iron around his head and [to always return] to camp with nausea.[2]

Desmond goes on to point out that Hooker was not well at the time, having a heart condition that got him 'excited over trifles'.

Hooker's observations were only the first to cast some doubt on the Kabru ascent because Graham had claimed the altitude not to be a hindrance. The doubters largely lost that argument when, in 1907, on Trisul (7,122m), Tom Longstaff and his team climbed over 1,800m from their highest camp at 5,300m to the summit in a day.

The next team to tackle Kabru, in 1907, were two Norwegians: Ingvald Monrad Aas, who had never climbed before, and Carl Rubenson, who had climbed in Norway. In his report in the *Alpine Journal*, Rubenson states that they 'did not suffer any real physical inconveniences with the altitude'. As for Hooker, Rubenson's reservations were withdrawn, 'but Mr Longstaff on his last expedition proved that such rapid progress was not impossible, and

---

1   *Alpine Journal*, Volume XII, August 1884.
2   *Proceedings of the Royal Geographical Society*, Volume VI, 1884.

I do not venture to dispute Graham's statement any longer'. (The Norwegians' expedition is discussed in greater detail in Chapter 9.)

The dispute was kept alive by those with axes to grind. Sir Martin Conway had made an ascent in 1892 of Pioneer Peak in the Karakoram with C.G. Bruce, Mattias Zurbriggen and two Gurkhas, reaching a height of just over 7,000m. Using others' spurious arguments, Conway, in his contribution to *The Encyclopaedia of Sport* in 1898, dismissed Graham's account, claiming that he held the record from his ascent on Pioneer Peak. However, in the edition of the *Encyclopaedia* in 1911, he evidently had had a rethink and gave Graham full credit for having reached the highest altitude at that time.

The American William Hunter Workman wrote a paper for the *Alpine Journal* of August 1905, on 'Some Obstacles to Himalayan Mountaineering and the History of a Record Ascent'. In his article, he claims to hold the height record from having climbed Pyramid Peak (now more usually Spantik, 7,027m) in the Karakoram in 1903. In a footnote he considered Graham's claim irrelevant as it had been:

> so strongly disputed that it must be regarded as far from proved, and therefore the altitude mentioned cannot properly claim a place among those acknowledged to have been made.

Workman did not state categorically that Graham had not climbed Kabru but preferred rather disingenuously to suggest that Graham's claim was irrelevant because it could not be proven objectively.

The officers of the Indian Survey would appear to have been self-serving: they closed ranks when criticised. In his address to the Royal Geographical Society and subsequently, Graham and Boss were highly critical of the quality of the cartography in the Garhwal. Emil Boss, the day after the Royal Geographical Society presentation, spoke at the Alpine Club where he complemented Graham's account, highlighting how they were so lucky to have perfect snow conditions on that critical point of the climb. He then elaborated on Graham's comments and criticisms of their Survey of India maps and how they might be remedied. His speech was by no means all negative: he was full of admiration for the Great Trigonometrical Survey. He did, however, reaffirm their earlier suggestion that Survey of India officers should be sent to Switzerland for training in ice-craft and cartography in glaciated regions. Douglas Freshfield, who was in the audience, endorsed everything that Boss had said and later wrote in *Round Kangchenjunga* of Graham's ascent:

> Much of the criticism bestowed on it has arisen from crass ignorance of mountaineering.

Freshfield later wrote, after talks with Survey officials:

> The result of this enquiry was to convince me that the opinion of officials, none of whom had any knowledge of mountain-craft, was worth no more in the Himalaya than it was in the Caucasus or elsewhere.

Graham and Boss's criticisms were intended to be helpful. Unfortunately, the Survey of India officers saw the climbers' criticism as anything but constructive.

Any student of this controversy will be indebted to Willy Blaser and Glyn Hughes for their reassessment, including their forensic scrutiny of the comments pouring scorn on Graham's ascent which appeared in the *Pioneer Mail, Allahabad* on 27 July 1884. The article was written anonymously by 'one who has been nearly thirty years a wanderer in the Himalaya', subsequently designated 'Wanderer'. Blaser and Hughes convincingly dismiss all the Wanderer's arguments.

However, one surveyor who had to be taken seriously was Kenneth Mason, former Superintendent of the Survey of India and author of the very important *Abode of Snow* (1955). He categorically stated:

> My own considered opinion as a mountain surveyor for what it is worth is that Graham did not climb Kabru.

He then cites in support the opinion of Tanner and Waddell, surveyors but not mountaineers. Colonel Tanner's credentials for commenting on climbs above the treeline are somewhat suspect, in view of his support for Harman and Roberts's declaration that 'Kangchenjunga may be said to have no glaciers worthy of the name.' He also lists Conway's comments on his own ascent of Pioneer Peak as evidence against Graham, unaware, it seems, that Conway had changed his mind by 1909 and was now supporting Graham. Mason then foolishly named Freshfield, Collie, Garwood, Waddell, Longstaff and Rubenson, who 'have argued for or against the claims' but failed to inform his reader that only Waddell, by quoting Conway and the surveyors, was 'against', whereas all the others became firm supporters of Graham's Kabru ascent. Mason backed up his argument against Graham by saying that he had mistaken his mountain by failing to mention a glacier that Cooke had found so time-consuming on his ascent of Kabru in 1935. It appears Mason failed to read all of Graham's account thoroughly and therefore did not know that Graham had mentioned the glacier problem on his 6 September reconnaissance of the southern approaches to Kabru. It was obvious even to seasoned mountaineers at that time that the Kabru trio could not have mistaken their mountain. Norman Collie pointed out that they must have climbed Kabru, 'because there is no other high peak which he could have ascended from his starting point except Kangchenjunga itself'.

Longstaff, one of the most illustrious Himalayan climbers of that period, wrote in the *Alpine Journal*:

> One word more before we leave the subject of the greatest Himalayan expedition that has yet been made. Twenty years ago, strange ideas were prevalent even in this country on the subject of mountain-sickness, ideas which have not yet entirely disappeared. In India at the time such ideas were probably more exaggerated, and ignorance of mountaineering matters was almost universal. Furthermore, by an unreasonably severe criticism of the G.T.S., Graham had set the officials of the Survey Department against him. Thus, mainly from ignorance, most people in India refused to believe in his ascent of Kabru. A well-known Indian official of my acquaintance, who was at Darjeeling at the time of Graham's visit, says now, and said then, that he fully believed in Graham's bona fides, but thought that he had mistaken Kabur (15,830 feet) for Kabru (24,005 feet), an opinion which has since been quoted by others. Now, for anyone who is a mountaineer, and has seen Kabru, it is impossible to believe that Graham, Emil Boss, and Kaufmann could make any mistake as to what peak they were on. They may have been impostors, but they could not have been mistaken; my point is that we have no tittle of evidence to show that they were either. Any climber who will carefully study Graham's paper in its entirety, especially if he knows the country at all, cannot but be struck by the strong internal evidence of truth which it bears. That he did not suffer from mountain-sickness is no proof of bad faith. That he made little pretension to scientific knowledge is not evidence that he was not a very competent mountaineer.

However, Graham's account of his disputed ascent of Changabang in *From the Equator to the Pole* (1887), and in the *Geographical Journal* and *Alpine Journal*, in conjunction with the very detailed Swiss map *Garhwal Himalaya Ost*, all help to ascertain the probable peak he climbed. On the Swiss map, some two miles south-east of Dunagiri, two points of the same height, 6,140m, are marked less than 800 metres apart, which would suggest that this is the 'level top' peak Graham climbed. However, he does himself no favours by writing:

> We slept at about 18,000 feet and the next day achieved the ascent very successfully from the western ridge. It was a fair climb but presented no great difficulties.[3]

Giving inconsistent directions leaves Graham open to criticism, as well as confusing future climbers. Kenneth Mason suggested:

> It is more likely that the summit he reached was on the southern or Hanuman Ridge of Dunagiri, possibly that shown on the latest maps at 19,210 feet.

Although Graham is rather casual with his directions, south-south-west is substantively different from south-east!

So, having made one error, in mistaking Changabang for a point to the south-south-west of Dunagiri marked 6,077m and named 'Hanuman', we do not know exactly which peak Graham was referring to. It does not follow, though, that we should lose faith in *all* his directions; and it certainly does not mean that, having made one mistake, everything else he achieved should be ignored or written off as irrelevant. (In *Abode of Snow*, Appendix C provides a very useful 'Chronological Summary' of the main explorers and climbers in the Himalaya. Reference is made to Rubenson and Aas reaching 7,270m on Kabru in 1907 but no mention is made of Graham reaching a similar height, or higher, on Kabru in 1883.)[4]

It has been disappointing to find that the Royal Geographical Society publication *Mountains of the Gods* (1984) unambiguously takes the negative view that Graham was mistaken about Kabru:

> It now seems probable that the peak he climbed was not Kabru but the much lower Forked Peak (20,340 feet).

---

3 *Proceedings of the Royal Geographical Society*, Volume 6, 1884.
4 Some sources say 7,287m.

The authors themselves, however, express some disputable opinions in this work.

More recently, the authors of *Fallen Giants* (2008) discuss the 'basic improbability of it, Graham's description of the climb makes little topographical sense and neither the Indian survey nor the Alpine Club ever credited it ... '. Their opinion, however, derives largely from Walt Unsworth and Kenneth Mason.[5]

Alexander Kellas also entertained ambitions for the height record. Kellas pointed out to Collie that Graham could not have ascended Kabru, 'because he mistook Kabru for Kanchengjunga'. His biographers do not elaborate, although they must wonder what peak is being referred to. The biography *Prelude to Everest*, by Ian Mitchell and George Rodway (2011), reveals that Kellas was not a technical climber. On the other hand, being a chemist, he applied himself diligently to understanding how best to cope with altitude. (The many excursions Kellas made in Sikkim with local people are discussed in Chapter 10.)

Unfortunately, the authors of *Prelude to Everest* do not subject Kellas's comments about Graham to critical appraisal; in fact, just the opposite. They state:

> Graham's climb was never officially recorded to him. Victorian gentlemen did not stoop to accusing each other of falsehoods but the overwhelming rejection of Graham's application to join the Alpine Club speaks volumes.

But does it? Not if he was rejected before he ever went to the Himalaya and for something (as we have seen) quite separate from Kabru. They refer to Graham as 'possibly infamous', which is a defamation of Graham's character and, by association, that of the two highly regarded Swiss guides, Kaufmann and Boss. The authors, perhaps to elevate their man, declare, 'Kellas knew of the failure of the Norwegians, Rubenson and Aas on Kabru in 1907, but there remained the prior claim of Graham to have conquered Kabru back in the 1880s.'

Neither Graham nor Rubenson claimed they reached the summit of Kabru, so it registers as odd that the authors write up one as a failure and the other as having 'conquered' Kabru. (Kellas had attempted Kabru but turned back almost 1,000m below the summit.)

---

5   Editor's note: while preparing this book, I interviewed Professor Stewart Weaver, co-author of *Fallen Giants*, on this matter. His opinion is what is written.

Kellas's biographers introduce the idea that Graham intended to circumnavigate Kangchenjunga in 1883:

> Though Graham failed ... in the following year a Pundit ... Rimzing Namgyal did accomplish exactly that feat ...

Nowhere is it recorded that this was Graham's intention and, if it was, on what date did he set off and 'fail'? If there is a reprint of their book the authors might revise it to include assessments from the large number of contemporaneous mountaineers who came out in favour of Graham's ascent, and why. Any reader will want to trust an author to 'get it right', as I did when reading *Prelude to Everest* in order to write the foreword! The authors might also like to make the point that if Graham and his party had not achieved a height record, and if he had not been critical of the Survey of India, then there might not have been any controversy at all. In that case, their reports would have been accepted in the normal way, as were those of Alexander Kellas.

Graham and his team were fit and enthusiastic. After Kabru, they again crossed the Kang La into Nepal where they climbed a peak of nearly 5,790m, from which they could inspect another great mountain on their list, Jannu (7,710m). They decided it was now too late in the year to make an attempt and returned to Darjeeling. Even then, Graham and Boss had the energy to go hunting in the Terai. These are not the actions of men hooked on breaking records and rushing home for acclaim. What they did they evidently achieved for themselves, with no need to cheat or even embellish their successful climbing season.

William Graham, soon after his Himalayan season, disappeared from the climbing scene, but not before many mountaineers had endorsed his and his Swiss companions' achievements. At the time it was thought by Collie that Graham had lost money and departed for the Wild West as a cowboy! Walt Unsworth, however, in his superb book *Hold the Heights: the Foundations of Mountaineering* (1993), tells us that after writing to the Foreign Office on 4 July 1986, he was informed that:

> He [Graham] was actually British Vice-consul in Durango, a small town in Mexico from 1910–1932 when the post was closed. He would then have been about seventy-three and he again slipped from view.

If Graham had been more accurate and consistent in reporting his climbs and had he not been quite so vocal in his condemnation of the Survey of India maps above the treeline and had there been more generosity of spirit among mountaineers and Survey people, the ground-breaking climbs of Graham, Boss and Kaufmann would have been taken far more seriously at the time. In which case, many more alpinists would have felt encouraged to visit the Himalaya and apply to their climbing the same 'Alpine style' of total commitment than did Longstaff, Kellas and the Norwegians.

9

Crowley and the Norwegians

The first attempt to climb Kangchenjunga
The first expedition intending to reach the summit of Kangchenjunga left Darjeeling on 8 August 1905 and returned there from having reached a high point of around 6,400m.[1] One porter fell to his death; three other porters and a Swiss member of the climbing team all perished in an avalanche. Of the many expeditions to the Kangchenjunga massif, including William Graham's, this expedition proved to be the most notorious.

The prime mover and climbing leader was Aleister Crowley. His ill-matched team consisted of Dr Jules Jacot-Guillarmod, Charles Adolphe Reymond, Alexis Pache and Alcesti C. Rigo de Righi. Argument and discord began on the walk-in; it increased with gain in altitude. The outcome of this failure to reconcile differences was a mutiny against Crowley's leadership, followed immediately by four needless deaths. Crowley was held responsible at the time. Subsequently, some commentators have taken a more sympathetic view of those notorious events above the Yalung glacier.

Aleister Crowley remains the most infamous character in the history of British mountaineering. In some of his numerous biographies, however, he has been found to have some redeeming qualities.

During the summer of 1892, Crowley and his mother stayed at the Sligachan Inn on Skye. A party of climbers staying at the hotel took him up Sgùrr nan Gillean ('The Peak of the Young Men', 964m) by the Pinnacle Ridge, which he found testing but also a revelation.[2] He eventually put up some of the hardest and most exposed routes on Skye.

---

1   Editor's note: more up-to-date estimates suggest that c.6,000m might be nearer the mark.
2   Pinnacle Ridge of Sgùrr nan Gillean: graded now at Hard HD; perhaps slightly easier in Crowley's day because less polished (though with much more debris). The route includes considerable exposure and one overhang abseil.

In 1896, he climbed at Beachy Head, not far from his new home in Eastbourne, where he lived with an Exclusive Brethren tutor while attending occasional chemistry classes at Eastbourne College.[3] The more he became familiar with the 150m chalk cliffs, the more they became his passion. He wrote, 'The ordinary man looking at a mountain is like an illiterate person looking at a Greek manuscript.'

These and other climbs were incredibly bold achievements. It was, he declared:

> A matter of the most exquisite judgement to put on it no more weight than is necessary; a jerk or a spring would almost infallibly lead to disaster. One does not climb the cliffs, one hardly even crawls. Trickle or oozes would perhaps be the ideal verb.

He always led from the base of the climb to the top without prior inspection from a top rope.

Mick Fowler points out in *Vertical Pleasure* (1995) that Crowley is remembered by the public for:

> his black magic routine with which he regularly terrorised other devotees, but [by] climbers for his exploratory adventures ... light years ahead of his time to tackling vertical chalk cliffs ... the ground he covered was without doubt amongst the most technically difficult in Britain ... it was not until 1979–1980 that a small group of climbers began to appreciate the joys of this coast and take over where Crowley left off some eighty-five years before.

Crowley also pushed out the boundaries from 1894 to 1898 during his Alpine seasons: he made the first guideless ascent of the Mönch. He continued to be acknowledged and admired for the style and difficulty of his Alpine routes, often with his mentor, Oscar Eckenstein.

He was a constant critic of the Alpine Club. Of its membership he wrote:

> all the people I met were constantly on the brink of disaster ... various old fogeys ... had been personally conducted by peasants up a few mountains and then written themselves up into fame ...

---

3   Exclusive Brethren: a splinter group of the Christian fundamentalist Plymouth Brethren.

they tried to ignore English rock-climbing altogether and would have nothing to do with the continental Alpine Clubs.

Crowley made the more serious accusation that some Alpine Club members falsified their records in the Alps. He accused Conway of only reaching the foot of Pioneer Peak in the Karakoram, according to Balti porters he had quizzed during his 1902 K2 expedition. Not only was he barred from joining the Alpine Club but the majority of members built up a loathing of him which in some cases persisted certainly into the 1950s and perhaps later. In a review of John Symonds's *The Great Beast: the life of Aleister Crowley* (1951), the former editor of the *Alpine Journal* T.S. Blakeney found the book:

> tedious reading, being largely a succession of drugging, suicides, debauchery, starvation, misery of all sorts, without any relieving features ... not merely the sport of mountaineering, but the world at large, would have suffered no loss if Kangchenjunga had permanently effaced Crowley.

To harbour such hatred of Crowley so long after his death indicates that the old guard at the Club defended themselves by attacking Crowley on aspects of his life other than climbing. A more reasoned approach would have been to examine the criticisms Crowley had raised, such as reactionary views concerning guideless climbers, pretentiousness, exaggeration and falsification of records.

Not all establishment figures of the Alpine Club were so intolerant of Crowley and Eckenstein. Tom Longstaff, in his splendid autobiography, *This My Voyage* (1950), looks back to August 1899:

> August 24 was an off day, beguiled by lessons in the use of crampons from that enigmatic couple Oscar Eckenstein and Aleister Crowley. Eckenstein was responsible for the introduction of ice-claws, crampons, amongst British mountaineers: he even claimed that their use made the regulation ice axe unnecessary. There was a peculiarly English opposition to their use in those days; they were scornfully called 'artificial aids' but surely, so are special nails in climbing boots or an ice axe.
>
> ... as Laird of Boleskine, [Crowley] chose to wear full Highland

dress, complete with eagle's feather, on all occasions.[4] ... I have seen him go up the dangerous and difficult right (true) side of the great icefall of the Mer de Glace below the Géant alone, just for a promenade. Probably the first and perhaps the only time this mad, dangerous route had been taken.

By 1898, Crowley had begun to climb more often with Oscar Eckenstein, another mountaineer who had alienated himself from the Alpine Club. Eckenstein had been forced to leave the Conway Karakoram expedition without the opportunity of visiting Concordia and K2. Thereafter, climbing K2 became something of an obsession. He therefore proposed to Crowley that they put an expedition together to climb K2 in 1902.

On returning from K2, and feeling lost, Crowley married Rose Kelly, with whom he had a child. During the unravelling of his marriage, in which it seems there was genuine love for each other, Crowley decided to climb the world's third-highest peak, Kangchenjunga.

The Swiss army doctor on the K2 expedition, Jules Jacot-Guillarmod, had put the idea forward when visiting Crowley at Boleskine House in 1904. Crowley had little respect for the doctor, finding him amusing but thinking 'he knew as little of mountains as he did of medicine'.

When the first expedition to attempt to climb Kangchenjunga left Darjeeling, its team was already fatally flawed; for that, Crowley, with typical honesty, blamed himself. Oscar Eckenstein refused Crowley's invitation as he felt Guillarmod's membership would lead to disaster. To minimise any risks, Crowley and Guillarmod drew up a contract. The team had to sign it, acknowledging that Crowley would be giving the orders on the mountain and had to be obeyed. With that flimsy paper guarantee they set off because, as Crowley wrote:

> I liked Tartarin so well, personally, that I unconsciously minimised his imbecility.[5]

---

[4] Crowley had bought Boleskine (pronounced 'bo–LESS–kin', with accent on the second syllable) House, above the east bank of Loch Ness, just north of Foyers. It subsequently burnt down but has recently been bought and is being renovated. His title 'Laird' is pretentious: it should only be used by the owner of a long-established estate. 'Laird' is perceived as being 'higher' than a gentleman but 'lower' than a baron, a status which does not square with Crowley's background. The eagle's feather is another piece of pretension, it being reserved for clan chiefs only.

[5] Tartarin is probably an allusion to the eponymous hero of Alphonse Daudet's 1872 novel *Tartarin de Tarascon*, about a gullible traveller who is beset by frequent misadventures.

The omens were not good: on 15 August, 'I had a telegram from the doctor that he had been ship-wrecked in the Red Sea. I might have known it!' When Guillarmod did arrive:

> The doctor seemed to be suffering from ill-health from various trifling causes. He seemed a shade irritable and fussy. I suspect the cause was partly physical. His sense of his own importance was hurting him.

Another ill-matched member of the team, for whose awkwardness Crowley also took full responsibility, was an Italian, Alcesti C. Rigo de Righi, the young manager of the Drum Druid Hotel in Darjeeling. He offered to join the expedition as transport officer and so Crowley:

> relying on his knowledge of the language and the natives, thought it best to accept him, though his character was mean and suspicious and his sense of inferiority ... manifested itself as a mixture of servility and insolence ... and of swaggering and bullying to the natives. These traits did not seem so important in Darjeeling, but I must blame myself for not foreseeing that his pin brain would entirely give way as soon as he got out of the world of waiters.

The remaining team members introduced by Guillarmod were both Swiss: Lieutenant Alexis Pache, an army officer, and Charles Reymond, who had completed several guideless climbs in the Alps. Crowley also brought from the north-west of India three Kashmiris who had been stalwarts of the K2 expedition: Salama, Subhana and Ramzana. There were also six personal servants, the sirdar and seventy-nine 'regular coolies'.

Crowley had:

> reconnoitred Kangchenjunga from England, thanks to the admirable photographs of every side of the mountain taken by Signor Vittorio Sella, who accompanied some man named Freshfield in a sort of old-world tour around the mountain. I had also a map by Professor Garwood, the only trouble with which was that, not having been up the Yalung glacier himself, he had had to fill in the details from what he himself calls the unintelligible hieroglyphics of a native surveyor, who had not been there either.

Crowley took the precaution of sending '8,000 pounds of food for the coolies to a depot as near the Yalung glacier as possible' with the government transport officer, Major White. The Major lost control of his porters, who according to Crowley had:

> no notion of self-respect, no loyalty, no honesty and no courage. Many of Major White's men deserted, either dumping their loads anywhere on the way or stealing them; and there was no means of controlling their actions.

What did Crowley make of the journey from Darjeeling to the unexplored Yalung glacier and Base Camp? Of Darjeeling he was not a fan:

> Sir Joseph Hooker, one of the few men of brains who have explored these parts, made an extended survey of the district and recommended Chumbi as a hill station. 'Oh well,' they say, 'Darjeeling is forty miles nearer than Chumbi. It will do rather better.' So, Darjeeling it was. The difference happens to be that Chumbi has a rainfall of some forty inches a year; Darjeeling some two hundred odd ... The whole town stinks of mildew. One's room is covered with mildew afresh every morning ... The food itself is as mildewed as the maidens. The hotels extort outrageous rates which they attempt to justify by describing the meals in bad French. To be reminded of Paillard is adding insult to injury, for what the dishes are made of I never did discover.[6] Almost the whole time I was there I was suffering from sore throat, arthritis, every plague that pertains to chronic soddenness. Do I like Darjeeling? I do not!

The route north from Darjeeling towards Kangchenjunga, over the fifty-five years since Joseph Hooker's visit, had been largely tamed with, according to Crowley, an excellent ridgeway path passing through Jorpakr, Tonghu, Sandakphu, Phalut and Chabanjong, with the first four stages having the benefit of 'well-favoured dak baghlas' (bungalows). Two unpleasant aspects to the journey surfaced: the rain and the leeches. Crowley said he thought he knew all about rain. Only after five days of constant downpour did it slacken.

The route becomes vague in Crowley's narrative owing to a confusion of names, particularly when they could no longer keep to the crest of the

---

6   Paillard: tenderised meat.

Singalila Ridge. On 18 September, after an eight-hour walking day, including descending some 900 metres into a valley, they camped at Gamotang. This descent, he wrote:

> was like stepping into fairyland. There are certain places ... which possess the quality of soft brilliance quite unearthly – they stand out; as a genius does from a hundred other men in evening dress.

From Gamotang they climbed in four hours to the Chumbab La (4,520m). It seems that they then skirted the west side of the Kang La to join the track from Kang La to Tseram. They camped outside Tseram, which consists of just a few huts by the Yalung Chu. Crowley's general principles required him to have nothing to do with the natives:

> The Dewan of Nepal was sending an officer to superintend our journey. I should perhaps have mentioned before that England has a special treaty with Nepal, one of its terms to the effect that no foreigners are to enter the Dewan's dominions. He knew that the most harmless of Europeans is the herald of disaster to any independent country. Where the white man sets his foot, the grass of freedom and the flower of good faith are trampled into the mire of vice and commercialism.

On 21 September, Crowley left Pache in charge of Camp 1 just upstream from Tseram. Now among the mountains, he proceeded with a small reconnaissance group on a good track up a valley, 'gorgeously wild, it glowed with rich bright grass and masses of marvellous flowers', as it does on a fine day after heavy rain. He passed under the huge flanks of the Rathong Peak (6,682m) and main Kabru summit (7,412m), although Kangchenjunga itself remained hidden in clouds. The walk to Camp 2 was easy, he says, 'with the summit of Kangchenjunga ... only two miles away'. That would situate Camp 2 well past the modern Base Camp and halfway up the south-west face. As this is impossible, we must question all the heights and position of further camps. Crowley does say he cannot pretend to any accuracy about either height or distance as:

> Tartarin was in charge of the surveying and I do not know whether he took proper observations at all. All I can say is that Professor Garwood's map is seriously wrong in many important points.

Crowley now walked on to Camp 3, with nearly all the rest of the expedition strung out below him as far as Camp 1. Many 'coolies' had deserted the expedition; it seems that had it not been for Pache, many more would have left because of Righi's attitude towards them. Pache managed 'to get them into good humour again', taking charge 'as soon as he saw that Righi was half insane with the fear that come to people of his class in the absence of a chattering herd of his fellows'.

Yet, despite Crowley having sent a man down to extol the virtues of the Camp 3 site, Guillarmod obliged the main body of porters to camp on bare ice, less than an hour away.

Next day, according to Crowley, he had soon brought all the porters up to Camp 3, to the amazement of 'Tartarin':

> He claimed that they had positively refused to go further, but they picked up their loads and strolled cheerfully up the slopes without so much as a word of admonition. Not understanding the secret of my power, though he had seen it exercised so often by Eckenstein and myself, he imagined that I must be terrorising the men by threats and beatings. In point of fact, I never struck a man during the entire expedition, save on one occasion to be described presently.

Camp 3 was estimated to lie at 5,500m; it afforded magnificent views in every direction. It was the haunt of small birds, which probably puts it at Yalung Base Camp (5,402m) as marked on modern maps.

Crowley continued the ascent, to establish Camp 4 with Guillarmod, who was distressed:

> but recovered by the next morning sufficiently to curse. I could not imagine what his grievance was ... The most charitable explanation ... is that he was mentally upset ... due to physical distress, and to some form of heat stroke.

These frequent problems of communication occurred either, as Crowley suggests, out of 'absolute perversity' or possibly because Guillarmod and to some extent Reymond had lost respect for the leader. Crowley never doubts his role in the misunderstandings which permeated this expedition. It is very tempting to be out in front, but it is not the place from which to direct operations or to 'manage' an expedition properly: Crowley preferred to label himself 'manager' rather than leader.

Suddenly, far too many porters had arrived at Camp 4. Crowley had intended Guillarmod to billet them at the more commodious Camp 3. Word came up with Pache that there had been desertions; one man going down alone had perished in a fall. Crowley sent the doctor down to make enquiries and to send up food and fuel that Righi was refusing to release. On 19 August, Reymond, Pache and Crowley ascended with a small party and reached Camp 5 in about three hours. Crowley established the height of this camp at between 6,100m and 6,400m (see note 1).

On 31 August, Crowley plus six men, including Salama and Reymond, climbed above Camp 5, clearing loose snow and ice and making bucket steps for the porters. Ice chips rained down, starting a minor avalanche which unnerved the porter Gali, who became hysterical until Crowley whacked him with his ice axe.

They retreated safely to Camp 5 but the porters:

> began to talk nonsense about the demons of Kangchenjunga and magnified the toy avalanche and Gali's slip and wallop to the wildest fantasies. During the night some of them slipped away and went down to Camp 3.

Reymond, Pache and Salama, on 1 September, pushed the route out by fixing a rope over a bad section of ice. At Camp 5, some seventeen to twenty 'coolies' had arrived with Guillarmod and Righi. They announced that they had held a meeting below and decided Guillarmod should take over the leadership of the expedition. For the time being Crowley brushed aside their mutiny, realising that the immediate problems were lack of room at Camp 5 and the deteriorating snow conditions of late afternoon. He sent the porters down. They safely arrived at Camp 4, but he implored Guillarmod, Righi and also Pache, who had now sided with Guillarmod, not to descend. Despite an impassioned plea to Pache, he still went down with Guillarmod, Righi and four porters. He summed up the situation:

> Indeed, I had much more to worry me than the nonsense of Tartarin and Righi. They had brought up all these men without any provisions for food or shelter and it was now late in the day. The snow was in an absolutely unsafe condition and though I had chosen the route so as to minimise the danger, it was absolutely criminal to send me down. But the mutineers were utterly insensible to the voice of reason. I told the coolies that since they could not stay at

Camp 5, the best thing they could do was to shelter under the rocks at Camp 4, and they went off and did so. I warned the mutineers that they would certainly be killed if they tried to go down that night; it was perhaps more or less right for coolies, but for THEM – I knew only too well the extent of Tartarin's ingenuity in producing accidents out of the most apparently unpromising material. They stormed all the more. I ought to have broken the doctor's leg with an axe, but I was too young to take such a responsibility. It would have been hard to prove afterwards that I had saved him by so doing.

Having heard frantic cries from Guillarmod and Righi, Reymond, who still had his boots on, went out to investigate, with a request from Crowley to call him if he was needed. Since there was no word from Reymond, Crowley went to sleep. Early next morning, he descended to find that five of the team were now dead. Righi later explained that Guillarmod was out in front, with everyone else tied to a short rope behind, including the three porters who came last. Two of them, at the top of the slope, slipped off, pulling everyone else down and creating an avalanche in which they were buried, along with Pache. Guillarmod and Righi escaped with minor bruises. Crowley saw to the burial of Pache, while the porters placed their own people ceremonially in a crevasse.

The expedition was over. The members made their various ways back to Darjeeling and home, to confront or revel in the various accounts they had written of the expedition in local Indian and national British newspapers.

Crowley, because of his reputation and provocative accounts, attracted most of the blame for the accidents. Was this justified? This tragedy could easily have been avoided if Guillarmod, Righi and Pache had stayed at Camp 5 as Crowley strongly advised, because of the state of the snow. The accident might have been stopped in its tracks if the experienced doctor and Pache had been in the rear instead of out front, so that as the two porters slipped, they could immediately have checked their fall.

Subsequently, Crowley came in for more criticism for not going out on hearing the commotion from Guillarmod and Righi. However, after a hard day on the hill, having laboriously taken off boots and clothing and crawled into his sleeping bag, it was to some extent reasonable to leave it to 'fully dressed and shod' Reymond to check the situation. Crowley did after all ask him to call if needed, which he did not.

Crowley did nothing to avoid criticism, and indeed encouraged it by his articles appearing in the *India Pioneer* which were widely reported across the world. Writing the following did not help his cause, unless of course he was relishing gaining further notoriety:

> I could do nothing more than send out Reymond on the forlorn hope. Not that I was overanxious in the circumstances to render help. A mountain 'accident' of this sort is one of the things for which I have no sympathy whatever.

In another article he consoled himself while lying in his tent with the thought that:

> the doctor is old enough to rescue himself and nobody would want to rescue Righi.

Guillarmod was, as Robin Campbell points out, 'a fairly ineffectual character'. One of Guillarmod's ridiculous ideas was to object to early starts. Crowley wanted to go off at dawn to complete the day's climb before the sun made the snow unsafe. Guillarmod thought it cruel to expose the porters to the cold. Crowley, on the other hand, having been designated the leader, should have ensured the high-altitude porters were well shod and provided with some form of crampons, or nails and ice axe. No apology for or explanation as to why the porters were so neglected in this matter was forthcoming.

Crowley never wavered from the views he expressed about the accident. His justification rested on his basic belief that everyone should take personal responsibility for their actions. 'Do what thou wilt shall be the whole of the law' was his credo, the distillation of his outlook and the reason that he did not feel responsible for the actions of Guillarmod and Pache, who alone made the decision to descend so late in the day.[7]

Alan Hankinson sums it up:

> Perhaps in the end Crowley's great crime in the eyes of his contemporaries was not so much his behaviour on the mountain as his refusal to apologise or find some excuse for it or even to shut up about it. This was, after all, the golden age of British hypocrisy – of canting piety and moral prudery and mealy-mouthed pomposity. Crowley's whole life was a violent reaction against this. He hated

---

7   'Do what thou wilt … ' was an axiom of Thelema, the esoteric spiritual philosophy founded by Crowley.

almost everything about late Victorian and Edwardian society, its chauvinism and racism, its mindless respect for class and money, its exploitations and extortions, its puritanical denial of the needs of man's nature, the jobbery and corruption of the 'old boy network', the superior public school attitudes and the belief that it was somehow vulgar to work too hard at anything.

In fact, the more you look at him the more Crowley emerges as a man of our time rather than his own. Had he been born sixty years later he might have enjoyed great success as a campus guru in California or a prickly pundit on television. He would almost certainly have been spared the slow decline into disrepute and mystical dottiness and litigation and seediness that was his lot. For the basic tenets of his belief – 'Do what thou wilt', stretch yourself both physically and spiritually, experiment with experiences, defy convention – accord closely with those of Western youth in the 1960s and '70s. He had intellectual vigour and considerable moral courage and a fine command of English prose. And he was not boring. The accounts of his expeditions in *Confessions* are as vivid and entertaining as anything ever written in that line.'[8]

After 1905, there is no record of Crowley climbing again, although he tried in vain to gather support in America for a 1906 attempt on Kangchenjunga. Robin Campbell writes:

It seems likely that Crowley's acid pen, whose prose output was in my view his best claim to genius, was alone responsible for his fall from mountaineering favour.[9]

What did Crowley and his team achieve on Kangchenjunga? The expedition had shown the way to the south-west face of Kangchenjunga for the first time and, thanks to Crowley's planning and leadership, had arrived at Camp 3 by the Yalung glacier within two weeks of leaving Darjeeling, despite heavy rain and undisciplined porters. The information on the vagaries of the weather would be of help in planning future expeditions, which could not now rule out the post-monsoon period.

The expedition also brought back useful information about the Yalung Valley and glacier, to be included in later editions of Garwood's map. They

---

8   *Climber & Rambler*, November 1979.
9   *Mountain*, No. 11, September 1970.

had paved the way for the first ascent of the mountain. Crowley and his friends probably reached a height of around 6,550m, certainly no higher than 6,700m (see note 1), and in doing so gave others the confidence to climb to such heights; in fact, Crowley claimed they were usually 'in excellent spirits' on the higher ground.

The accident was a setback, giving the south-west route up the mountain an extension of its reputation for avalanches, but Crowley's common sense had shown that measures could be taken to deal with adverse snow conditions and, just as he had done on K2, he identified the way by which the mountain was to be first climbed fifty years later.

The Norwegians on Kabru, 1907

On 20 October 1907, two Norwegian climbers, Carl Rubenson and Ingvald Monrad Aas, ascended to about thirty (vertical) metres from the northeastern summit of Kabru (7,338m). To have reached approximately 7,310m without European guides was a milestone in Himalayan mountaineering. This outstanding achievement astounded and inspired the climbing world: these climbers were aged only twenty and twenty-one, yet had reached that height by a new route. Reaching the summit area of this prominent peak to the south of Kangchenjunga was important to the progress of Himalayan mountaineering: the Norwegians, echoing W.W. Graham's comment, coped well with the altitude in reaching what was then an altitude record.

Carl Rubenson was the prime mover. He was born in 1885 in Stockholm to a Norwegian mother and Swedish father who moved to Oslo the year after his birth. On coming of age, his family settled on him a stipend sufficient to free him from further financial worries.

He set off for the Himalaya in the autumn of 1906 with his friend Ingvald Monrad Aas, who was just a few months younger. He was very fit but with no mountaineering experience. In Rubenson's opinion, however, he was a natural alpinist.

Many restrictions were imposed by the British on travellers through the north of India, but road and rail connections were progressively improving. Of all the Himalayan regions, Sikkim was easiest of access, politically and practically, at least as far as Darjeeling and Sikkim's capital, Gangtok.

Having decided not to employ Alpine guides, the Norwegians hired local people in Darjeeling to help carry loads. They reported the best of them were Sherpas, an ethnic group originating from the Khumbu region of Nepal. Sherpas were natural mountaineers. This chance encounter and awareness of the Sherpas' prowess at altitude led to them being employed on all the major expeditions to the highest peaks, including on the first ascent of Kangchenjunga.

Having decided to concentrate on climbing Kabru, they set out from the British hill station of Darjeeling, to head north into Sikkim to the pasture of Jongri (Dzongri) at about 3,950m. They climbed Kabur (4,826m), from where they had an overview of the glacier leading up to the south-eastern side of Kabru. This visit was made in February, out of season. They realised their best course of action would be to postpone their attempt on Kabru until early autumn. The summer months of 1907 were spent travelling round the Far East, before returning to Darjeeling in September.

With the help of Mr Mason, a Scot who had acted as interpreter on their reconnaissance, they gathered a hundred porters who were sent off to Yuksom on 15 September. The Norwegians and Mr Mason set off two days later, on horseback for the first three days, with a further two days on foot to Yuksom, the last village along the route to their mountain. It was there the heavens opened and the emerging leeches dined on the visitors. After sending back fifty of the porters, they continued north to the pasture of Jongri where they made camp and purchased many sheep for the remaining porters.

With the monsoon over and in better weather, they proceeded up the Rathong Chu on 6 October to gain the East Rathong glacier. It was here, beyond where supplies of wood could be obtained, that they discovered the porters had stolen most of their reserves of methylated spirits during the walk-in. To hide their crime, they had topped up the fuel cans with water, making what remained of the cooking fuel unusable. For the rest of the climb the Norwegians were now dependent on wood which had to be carried up to the highest camps.

From their idyllic base camp by a small tarn at 4,878m, next to a large rock on which they kept fit by bouldering, they moved out on 7 October. They took fourteen of their best men on to the glacier, skirting under Forked Peak and Kabru Dome to follow moraine at the base of their western flanks. They established Camp 2 at 5,500m and Camp 3 at 5,950m on the Kabru glacier, which descends from the east face of Kabru to the Rathong. Five nights were spent at this camp, from where the route through the heavily crevassed and broken ice of the Kabru glacier was prepared. Finally, they reached easier ground and established Camp 4 at 6,550m. They attempted to climb to the north-east summit of Kabru but gave up in the face of bitter cold and strong wind. Temperatures descended to -20 °C during their two nights at this camp. They decided to make another light camp (Camp 5) at 6,700m.

The porters were sent down to the lower camp, except for two who were more experienced, having been with the 1905 Kangchenjunga expedition.

On 20 October, the pair set off with one Sherpa for a final attempt on the summit. The Sherpa, complaining of hunger, turned back. Rubenson and Monrad Aas continued into the teeth of the 'icy cold wind blowing from the west, making it very unpleasant and difficult to walk'. They cut steps in the hard, wind-blasted snow and slowly gained height, changing direction to find what shelter they could from the wind. With darkness fast approaching, they were forced to climb steeper ground over rock and ice for two hours. At about 6 p.m. they reached:

> what we thought was the top, and then saw a low snow ridge some 50 or 60 feet higher, which I am certain was the actual summit. The sun had by this time set, and, having a rather difficult and dangerous descent, we did not dare go on, although had time and wind allowed us, this ridge did not present any difficulty and we could easily have climbed it. With regard to breathing we did not have any special difficulty. The cold was now almost unbearable.

Their prudence was justified: they experienced a narrow escape on the return to Camp 5 when Rubenson, coming down last, slipped and shot at great speed past Monrad Aas, who was able to arrest his fall by holding the rope as he lay against the slope. Four of the five strands of the rope had parted, so a fall of several thousand feet was averted by that single strand.

The Westerners and their Sherpa companions returned to Jongri on 21 October, having spent more than a fortnight on the ice at and above 5,950m. The Norwegians praised the porters especially, indicating it was thanks to them they were successful. The porters had been paid higher-than-usual rates and had been given good equipment and special rations of their choosing. Everything it seems was done to break down barriers and ensure all members of the expedition remained on good terms with each other, even to the extent of having organised games and sports at different camps.

On return to Calcutta, in his report, published in *The Englishman* of 4 November 1907, Rubenson wrote:

> The chief thing is to have as good and willing coolies as we had; properly fitted out and with kind treatment they will surmount what would seem impossible. Take it slowly and carefully, let the

coolies go over the road first without loads to get confidence, and they will then go with them. We could not make them use the rope when loaded; their reason for it, that they would not be able to help each other then, is worth considering. But by making as good steps as possible, bettering the road, fastening iron nails and stationary ropes on the most difficult places, we helped them as much as we could. Our experience is that the coolies, especially the Nepaulese Sherpa, are excellent men when treated properly, and our success is only due to the willingness and brave qualities of these people.

10

Kellas, Raeburn and Crawford

Alexander Kellas (1868-1921)

Kellas explored the mountains of Sikkim on six occasions between 1907 and 1921. His underlying motivation, according to his biography, *Prelude to Everest* (2011), by Ian Mitchell and George Rodway, was attempting to climb Kangchenjunga. In doing so, he became *the* expert on the effect of high altitude on human physiology. He was noted for encouraging others to employ Sherpas, as he had done on his Himalayan ascents.

Kellas had resigned from lecturing in chemistry at the Middlesex Hospital Medical School in 1919. Following a rest at the family home in Aberdeen he returned in April of the following year with his telephoto lens to Kamet and later to the Kang La to photograph Everest and continue to press himself to the limit of endurance. After his final visit to Sikkim, he travelled from there into Tibet as a member of the first Everest expedition. By then he was, at almost fifty-three, not a young man, nor was he well. He was so weakened by dysentery that he had to be carried on a stretcher across the wind-blown plateau at almost 5,000m. He made certain he stayed at the rear of the caravan as he could not bear the rest of the team seeing him in such distress. Just short of Kampa Dzong, he died from heart failure brought on by the dysentery and the strain he had put on his body during the previous two years while climbing on Kamet and Kabru. He was buried on 6 June 1921, on a hillside to the south of Kampa Dzong facing the peaks he had climbed over the border in Sikkim: Pauhunri, Chomiomo and Kangchenjhau.

Kellas was born in Aberdeen. Growing up, he spent most of his spare time walking in the Cairngorm mountains. In 1885, aged seventeen, he walked fifty-five kilometres from Ballater to the Shelter Stone with his younger brother, Henry, in twelve hours, excluding one hour's rest.[1] He had enormous stamina, being able to keep going in the Highlands of Scotland with inadequate food, clothing and shelter for days at a time, a characteristic of his later Himalayan explorations.

His was often a lonely existence, since he appears to have been socially alienated by inner voices leading to quite severe psychosis in later life. In the Himalaya he could find relief and be at peace with himself in the company of local people.

On his first visit in 1907, Kellas spent the first half of August crossing the Pir Panjal range in Kashmir. He then travelled by train to Darjeeling where he was joined by two Swiss guides, whom he does not identify. They made three attempts to climb Simvu (6,818m), in the Zemu glacier region of the Kangchenjunga massif, but failed due to deep snow. They also failed to reach the Nepal Gap (6,302m) between the Twins and Nepal Peak, owing to an impassable crevasse at around 5,800m. He expressed his hopes in a 1918 letter to Percy Farrar, President of the Alpine Club:

> I had expected to ascend Simvu (22,300 feet), and perhaps to attain 23,000 feet on Kanchenjunga.

In that same autumn of 1907, the Norwegians arrived to all but climb Kabru, with the help of local porters and guides. Kellas would have read their favourable comments on the Sherpa porters' high-altitude performance, as related in the *Alpine Journal* and *Geographical Journal*. Kellas was not, therefore, foremost in the promotion of the Sherpas for portering and climbing on Himalayan peaks; the Norwegians were first, but it was Kellas who really spread the message since, on his five subsequent visits to Sikkim, he wrote about the benefits he derived from their company. The Sherpas' ability to climb high, as suggested by Kellas, was demonstrated by the climbs achieved. Within a few years, and certainly before the first Everest expedition, Sherpas came to be preferred to the hill people from the Kumaon and Garhwal, or the Gurkhas.

In *Prelude to Everest*, the authors reveal the depth of mutual respect between this retiring Scottish explorer and his 'coolies'. (The use of 'coolie' may jar with today's reader who is so used to guardedly correct language. This word was in everyday use at the time, however, and rarely thought of

---

1 The Shelter Stone (Clach Dhian): a cave accommodating about six people formed by an enormous boulder resting on top of others. It is situated deep in the Cairngorms at the head of Loch Avon, at about 750m above sea level.

as derogatory. The porters themselves used 'coolie', recognising it was simply a fact of life that they were local native labour for hire.) Kellas appreciated his porters, especially Sona and Tuny. This was natural, since he was wholly reliant on local people for weeks at a time. Like the Norwegians on Kabru, he ensured his porters were adequately kitted out and well fed. He sometimes involved them in the planning and execution of his climbs. They, in turn, introduced Kellas to edible wild plants to supplement their diet and maybe, in the case of mountain rhubarb, to acclimatise. Kellas was impressed by the Sherpas coping so well in the Himalaya on a primarily vegetarian diet.

In 1909, Kellas visited Sikkim again. He was now a wiser man from his experiences of 1907 and from conducting scientific experiments on the effects of altitude on the human body, particularly on the changes to the red corpuscle count in the blood. He was to continue such experiments in the Himalaya until the year of his death.

He left Darjeeling on 7 August 1909 with sixty-two local porters and no European guides, but with Rigo de Righi, the Italian hotel manager from Darjeeling who had been the object of Crowley's derision during his 1905 Kangchenjunga attempt. Righi's adaptation to expedition life and high altitude had not improved: he failed to acclimatise and returned to Darjeeling after three weeks. Kellas otherwise found him a pleasant companion, and a good walker who obviously enjoyed being in the high mountains.

Their first objective was to cross the Donkia La in north-east Sikkim to attempt Pauhunri (7,128m). With Righi indisposed, Kellas, with two Sherpas, made an attempt but they were driven back by a snowstorm. They had camped at about 5,800m and reached just over 6,600m, from where they descended, to return later in the season.

Kellas performed rather better, perhaps after further acclimatisation, in the north-western sector of the kingdom, in the region of Kangchenjunga. His prime interest seems to have been to reach the highest summit ever climbed; the other was to explore Kangchenjunga. After camping on the Jonsong La (6,189m) on 5 September 1909, Kellas and his Sherpas explored the South Langpo glacier before descending to camp by the Kangchenjunga glacier at Pangpema. From there, on 9 September, they crossed the glacier to examine Kangchenjunga's north-west face.

Freshfield had implied there might be a summit route, via the glaciers that have their origins:

> in a snow plateau, or rather terrace, lying under the highest peak at an elevation of some 27,000 feet, that is only 1,200 feet below the summit ... this glacier affords what in my opinion is the only direct route to Kangchenjunga which is not impracticable.

What Kellas thought of this and how nearly he approached the ice cliffs is not known.

After his examination of the north-west face of Kangchenjunga, the party returned to the South Langpo glacier on 11 September. Two days later, they made an attempt on Langpo Peak (6,965m) from the Langpo Saddle (6,400m) but were driven back by a snowstorm. On 15 September, Kellas, with one Sherpa, reached the summit. The snow, he wrote later, was:

> in excellent order and the incident which pleased me best of all was the ascent [it being a first ascent].

The party re-crossed the Jonsong La on 17 September, to camp at the end of the Jonsong glacier. By 22 September they had gone up the South Lhonak glacier to the col at 6,555m, from where they ascended the North Buttress of Jonsong Peak to 6,700m, before giving up the attempt in mist and stormy weather. It would be another twenty-one years before Jonsong Peak was finally climbed.

With astounding persistence, Kellas then relocated his team to Green Lake via the Zemu glacier, from where he made another attempt at gaining Nepal Gap. As in 1907, he failed, this time owing to a four-day snowstorm depositing almost a metre of snow. The porters, who had been sent down for more food, could not return to resupply the camp. They therefore abandoned Green Lake and the Zemu glacier side of the massif, to traverse once again across northern Sikkim by way of the Lachen Chu Valley to Lhamo lake and Pauhunri. Kellas made a determined effort to climb it, but he and one Sherpa were stopped by deep snow and high, cold winds at just over 7,000m, only about 100m below the summit but over 500m higher than their 8 August attempt. This comprehensive visit to the north-east of Nepal demonstrated Kellas's enduring stamina as well as his underlying interest in ascending Kangchenjunga. Details of the trip are meagre, pared down to a diary of events as sent to Captain Farrar on 10 April 1919, ten years after his 1909 expedition to north Sikkim and published in the *Alpine Journal* only posthumously in May 1922.

Kellas's third visit to Sikkim was his most productive. In 1911 (not 1910 as stated in his Alpine Club obituary and then wrongly reported elsewhere), he and his porters came within fifteen vertical metres of the Nepal Gap. They then travelled north to the Chorten Nyima La to ascend, for the first time, Sentinel Peak (6,490m), to the east of the La. On this ascent, Tuny cut steps all the way to the summit.

The party then returned to Pauhunri, which they finally climbed on 17 June after a five-day battle against soft snow and high winds. At the time, Kellas believed he had ascended the second-highest peak in the world, after Longstaff's 1907 ascent of Trisul (7,120m). It was later discovered that Pauhunri was higher than Trisul; Kellas never knew that he held the record between 1911 and 1930 for having reached the highest summit. Had this achievement been known at the time, his interest in climbing Jongsong (formerly 'Jonsong') Peak (7,462m), to the north of Kangchenjunga, might have had a different perspective. In *Prelude to Everest*, the authors observe that:

> It is not to be discounted that Alec, shy and modest as he was, was purposefully aiming at Jonsong Peak as its ascent could have greatly raised his own profile amongst Himalayan mountaineers of his day.

In the event, Kellas judged conditions were against an ascent in 1911, before and after climbing Pauhunri with his two strongest Sherpas, Tuny and Sona.

They next turned their attention to Chumiomo (6,838m), on the main Himalayan divide, north-west of Gyaogang. After reconnoitring several possible approaches, in pouring rain, they moved to the north-west and, from a high camp, walked up snow to the summit on 12 July, returning to the camp by 4 p.m. (Kellas rounded off his 1911 visit to the Himalaya by taking his Sherpas to the Garhwal for a reconnaissance of Kamet, 7,744m.)

Kellas was a pioneer of many aspects of Himalayan climbing: each expedition was conducted in lightweight fashion, living mainly off the land and, after his first visit, relying on local porters not only as load-carriers but as climbing companions. He was forced into making economies since he did not possess private wealth to pursue what he considered to be 'the most philosophical sport in the world'. He went on to spend more time above 6,000m than any other climber of those days, on a shoestring budget.

In his paper 'The Mountains of Northern Sikkim and Garhwal', read to the Alpine Club on 6 February 1912 and published in the *Alpine Journal* in May 1912, Kellas mentions leaving Lachen on 24 April 1911:

with thirty-one coolies, eight of whom were Sherpa Nepalese who were to remain with us permanently, the remaining twenty-three being Lachen men who were to return after four days' march to the north west.

He concludes his paper by declaring:

> Many of the Sherpa Nepalese are first-rate climbers as well as coolies and could be used for serious climbing of the big peaks like Kangchenjunga, after proper training.

Kellas, with his scientific background as research chemist at the Middlesex Hospital Medical School in London, became increasingly interested in the physiology of extreme altitude. On each expedition, Kellas had incrementally accumulated personal experience of how the human body responds to the thin air, up to 7,100m on Pauhunri. After further experimental work near Dunagiri, Kellas finally left the Garhwal for Darjeeling on 9 November. He spent the winter in Darjeeling, having no family in Britain to await his return from Kamet. In February, he was officially invited to join the Everest expedition. During April, with four of his Sherpas, he climbed Narsing (5,801m), a southern outlier of the Talung glacier, and also reached the Kang La (c.4,973m), from where he took telephoto shots of Everest, to be used in *The Times* later in 1921 when reporting on the first Everest expedition. He returned to Kabru in the hope of squeezing in an ascent of (what he believed to be) the highest summit ever climbed, from where he planned to take informative photographs of Everest and its surrounding peaks. His other reason for a 'sustained attack on Kabru' was to train 'coolies' for the Mount Everest expedition.

The 1930 *Alpine Journal* reported:

> It is believed that he also reached the summit of Lama Anden (or Lamgebo, 19,250 feet/5,869m) about the same time.

(Joseph Hooker gave it the Lapcha name of Tukeham and a height of 5,937m. On the northern flanks, his 'floral finds' averaged ten a day: blue, pink and violet primulas and new rhododendrons.) Lama Anden is situated seven kilometres west of the village of Lachen and by 1921 was far more accessible to Kellas than it had been to Hooker.

Kellas ran out of time when 900m below the summit of Kabru. He departed to join the Everest reconnaissance party which left Darjeeling on 19 May 1921. By 31 May, Kellas was so unfit to ride the ponies that he had to be carried on a litter. On 5 June, he expired just as the expedition reached Kampa Dzong. The expedition lost the man who had more experience of high altitude than any other. As a result, the study of the physiological effects of high altitude that Kellas was to have undertaken, backed up by Professor J.S. Haldane and the Department of Scientific and Industrial Research, never took place.

He was unique as a Himalayan mountaineer. C.G. Bruce, in his *Himalayan Wanderer* (1934), acknowledges this, adding:

> But what an explorer he was and the most modest man that ever travelled the Himalaya.

The mountaineering and geographical publications at the time were full of praise for Kellas's unique contribution to Himalayan exploration. Yet Kellas hardly gained an acknowledgement in the 1950s, for example in Sir John Hunt's *The Ascent of Everest* (1953) or Charles Evans's *Kangchenjunga: the Untrodden Peak* (1956). The exception was Tom Longstaff, who was of Kellas's era and understood that Kellas should be set above all others, writing in *This My Voyage* (1950) that Kellas's death:

> was a very severe loss to Himalayan mountaineering: he had done more high climbing than any other man, and this with the sole assistance of Bhotias and Sherpas whom he had himself trained.

He goes on to recall that Kellas 'was a mountaineer of the utmost courage and resolution.'

Kellas's climbs were non-technical. In fact, he never took to rock-climbing in Britain and certainly avoided steep rock and ice in the Himalaya. Because of this, Percy Farrar, President of the Alpine Club and a member of the Everest Committee, opposed the suggestion by Douglas Freshfield that Kellas should be the 1921 expedition leader, since he held the view that 'Kellas has never climbed a mountain but has only walked about a steep snow slope with a lot of coolies, and the only time they got on a very steep place they all tumbled down and ought to have been killed.'

(The fall he refers to was on the descent of Kangchenjhau in August 1912.) This was unfair comment on Kellas: he was, after all, fifty-two at the time of his exemplary climbing on Kamet in 1920.

Kellas's reputation returned to prominence thanks to the high-altitude physiologist John West, of the University of California, San Diego. After several papers covering various aspects of Kellas's life in the *Journal of Applied Physiology*, West wrote a comprehensive article for the 1989–90 *Alpine Journal*: 'A M Kellas: Pioneer Himalayan Physiologist and Mountaineer'. In it, he recognises that Kellas's mountaineering experiences at high altitude were second to none. Since West is himself a scientist, he discusses Kellas's increasing interest and foresight in relation to the physiology of acclimatisation: in 1918, Kellas and the Oxford physiologist J.S. Haldane collaborated to develop their ideas about acclimatisation. They used the low-pressure chamber at the Lister Institute in London for three days. Their work together is described in *Suffer and Survive* (2007).

John West's greatest contribution to our understanding of Kellas's work lay in revealing the existence of an unpublished manuscript by Kellas from the summer of 1920, entitled 'A Consideration of the Possibility of Ascending Mount Everest'. This came about after Kellas received a request on 21 November 1919 from Captain T.E.C. Eaton, the Secretary of its British Section, to deliver a paper at the Alpine Congress of May 1920 in Monaco on 'The possibility of attaining an altitude of 8,800 metres'. The paper was translated into French and published in the *Proceedings of the Congrès de l'Alpinisme*.[2] West pointed out Kellas's conclusion that:

> Mount Everest could be ascended by a man of excellent physical and mental constitution in first-rate training, without adventitious aids if the physical difficulties of the mountain are not too great.

Fifty-eight years later he was proved correct when Habeler and Messner reached the summit in 1978 without bottled oxygen and in the following year, when the world's third-highest summit, Kangchenjunga, was climbed without supplementary oxygen being used at all on the mountain.

George Mallory described Kellas's burial to his friend and mentor, Geoffrey Winthrop Young:

> It was an extraordinarily affecting little ceremony burying Kellas on a stony hillside – a place on the edge of a great plain and looking across it to the three great snow peaks of his conquest. I shan't easily forget the four boys, his own trained mountainmen, children of

---

2   A copy of Kellas's original paper in English is lodged with the Royal Geographical Society, and a slightly later one, with amendments, is in the Alpine Club archives.

nature seated in wonder on a great stone near the grave while Bury read out the passage from the Corinthians.

Harold Andrew Raeburn (1865-1926)

Raeburn visited Kangchenjunga twice in 1920. On the first trip, he was accompanied by Lieutenant-Colonel H.W. Tobin, an experienced Himalayan traveller and mountaineer who later became editor of the *Himalayan Journal*. They left the British hill stations and a sandstorm on 22 July 1920 with a sirdar, cook and twenty-one porters with 'the object of the examination of the South East outliers of Kangchen, the investigation of possible (?) routes up its South East Face, and the complete traverse of the Talung Glacier', as he wrote in the *Alpine Journal*, November 1921.

Their journey, despite monsoon rains, leeches, bears and trackless jungle beyond Yuksom, improved the higher they walked, first through oaks, magnolias and chestnuts; then rhododendrons, pines and mountain bamboo; until finally, after crossing the western spurs of Jubonu (Jhopunnu, 6,524m), they spent a rest day among the alpine flowers at Alukthang (Onglakhang).

Almost half of their porters were then sent back, while the rest, with Colonel Tobin's encouragement, carried their heavy loads up the final 300 (vertical) metres of loose boulders to the Guicha (Goecha) La (4,940m). They then descended 600 metres to reach the grassy meadow known as Tongshyong Pertam. The weather on the morning of 3 August was fine, with clear views of 'Pandim's impossible north-western side' and of Kangchenjunga, of which Raeburn reported, 'All south eastern aspects of Kangchen's south peak are most repellent.' He also commented on the east ridge and the fact that it drops to a col before rising to Simvu (6,818m). Although Raeburn does not refer to it as such, the col is the Zemu Gap (5,884m), which he does acknowledge was reached by Kellas and his Sherpas in 1911, from the north. He thought Kellas had been wise not to descend the pass since the south side, as far as he could see, 'does not appear suitable for any but well trained and booted men'.

Raeburn and Tobin descended the Pandim side of the Talung glacier to its snout. They spent a week forcing a way down 'the wonderful gorge of the Talung or Rinpiram river'. After hacking through dripping forests and cutting steps up cliffs of clay with their ice axes, they came to a track and followed it to a clearing in the forest with burnt trees and patches of maize. The coolies cried with joy, 'Our lives are saved!'

They had finished the last of the rice that morning.[3]

---

3   Summarised from the *Alpine Journal*, November 1921.

Raeburn then reports:

A most picturesque aboriginal turned up. He was a Lepcha, dignified and taciturn, but guardedly friendly. His garment was a single sheet of cloth forming a kind of kilt, the upper part thrown over his right shoulder, leaving the left side of the body bare. At his side he wore the straight Lepcha 'ban' or knife, like a Roman sword. His colour was by no means dark, about that of old ivory, and features quite fine and far from unintelligent. The name of his settlement was, he told us, Tingla. It is not on any map.

Chief Tingla was evidently keen to pass on our hungry-looking company, and declared he had no food to spare, but that plenty was to be got at the village lower down the valley, called Sakyong. This is on the 1906 G.T.S. map of Sikhim. He sent a henchman to show us his bridge and the way ... [and] we found food in abundance. Our real difficulties here came to an end. From Sakyong we descended to the Talung and crossed by a long bridge of rotten rattans, supporting in V-shaped slings a couple of loose bamboo stems as roadway.

The crossing of this swinging, dancing, slack rope affair, over the leaping snowy cauldron of the great river, was a first experience to Tobin and myself. Perhaps neither of us enjoyed it, but of course we acted, for the coolies' benefit, as though we did. We passed Pontong on the other bank and stopped at Be, where is a Sikhimese bamboo rest-house. We spent a most interesting night in the Lepcha monastery of Lingthem, returning to Darjeeling via Gangtok and Teesta Bridge.

(Most of the route taken by Raeburn's party had been followed before, in 1900, by Sir Claude White and by Colonel Shawcross, neither of whom were serious mountaineers. They left little description of travelling in this part of Sikkim.)

Raeburn was one of the most experienced British climbers of that time; it was valuable to know his opinion on climbing Kangchenjunga from the south-east. He came from Edinburgh, where he followed his father by entering the brewing trade. He first began climbing in pursuit of birds' eggs: a part of his collection is kept in the National Museum in Edinburgh. He was known for his determination and physical strength and for putting up new routes throughout the Highlands of Scotland in summer and winter, especially on Ben Nevis. The first winter ascent of Observatory Ridge, at Easter 1920, just before heading off to Kangchenjunga, was his most important Nevis climb. With serious Alpine routes also climbed, as well as others in Norway and the Caucasus in 1913 and 1914, and after his mountaineering in the Kangchenjunga massif during the spring and the autumn of 1920, he was naturally considered for membership of the first Everest expedition.

Raeburn and C.G. Crawford (1890-1959)

For his second reconnaissance of Kangchenjunga, Raeburn obtained permission to enter Nepal to explore the mountain from the south-west. His companion was C.G. Crawford, who later became President of the Himalayan Club. They entered Nepal by the Semo La and crossed the snout of the Yalung glacier on 10 September. They descended for further acclimatisation to the winter village of Yengutang, from where they sent their now redundant porters back to Darjeeling. They returned to Tseram, where they left a few porters to relay wood up to the Yalung glacier camps. By 25 September they were ensconced at 'Tso' (lake) camp at 4,900m, after having crossed the side glacier (Tso glacier) which descended from Jannu.

They left Tso camp and some old tins and a boot-sole, the only relics of the 1905 (Crowley) expedition, for the east or left bank of the Yalung glacier. They camped at a sheltered green recess at 5,030m, called 'Nao', as a nearby herd of *bharal* were called that by the porters.[4] From here they diverged from the line Crowley had chosen in 1905 and tended east, to examine the aspect of Kangchenjunga facing the Talung Saddle.

On 28 September, with their sirdar Gyaljen and a few of the stronger porters, they crossed the glacier that descends from Talung Peak. They climbed the rock of the spur or rib that drops from Talung Peak to about 5,800m, before descending to a camp at 5,500m at the lower end of the Talung rib.

Next day, Raeburn, Crawford and Gyaljen reconnoitred the north side of the rib and camped at 5,650m on the snow-covered debris. Their aim was to reach the 'white mantle of snow which from Darjeeling may be seen to lie across the broad bosom of Kangchen, with a sickle-shaped gorget of rock at its upper extremity'.

On 30 September with Gyaljen and three 'booted coolies', they ascended the snow-covered branch of the Yalung flowing from the Talung Saddle and

---

4   *Bharal* (sometimes *burhal*): the Himalayan blue sheep.

made a camp at 6,100m after crossing to rocks on the face of Kangchenjunga. On 1 October, in sunny but very cold weather, Crawford, Gyaljen and Raeburn went up this right (north/east) flank of Kangchenjunga for about 300m, where they realised their experience and supplies were inadequate to go further. They spent one more night at 6,100m, then retreated, having observed that:

> The roar of the ice avalanches from Kangchen and Talung seldom ceased for long, day and night.

This lightweight visit constituted a useful step towards the mountain's first ascent. Charles Evans acknowledged that 'Raeburn, one of the most experienced mountaineers of his time', had picked out the likely route, 'by his experienced eye'.

(Aleister Crowley was not acknowledged.) Evans simply commented that 'Freshfield's shelf and Raeburn's white mantle of snow are now known as the Great Shelf ... and Raeburn's sickle-shaped gorget of rock [is] known as the Sickle.'

They left the area in two groups, with Gyaljen to accompany the camp and the majority of the porters to re-trace the walk-in route. On 4 October, Crawford, Raeburn and three porters, with two light tents, left for a col south of Rathong Peak in 'an attempt to convert this into a "La" to which the name Rathong might apply'. They made good time up the central moraine of the West Rathong glacier and had soon crossed the pass, to meet Gyaljen at Pemionchi monastery as arranged.

Raeburn was sent back from the Everest expedition, ill, to Sikkim, where he spent two months in hospital. He decided to return to the expedition, but his health, both physical and mental, went into general decline: towards the end, he believed he had caused Kellas's death. Five years after the Everest expedition, this great Scottish mountaineer died in his native city of Edinburgh.

11

Between the Wars I: British and American Expeditions

The two decades between the first and second world wars were among the most productive in the history of mountaineering in the Kangchenjunga massif.

1925: Nikolas Tombazi (1894-1986)

Nikolas Tombazi, born in 1894 at the Greek Embassy in St Petersburg, was a Greek photographer and geologist. He spent time in Sikkim in 1919 and 1920 and again in 1925, exploring the southern glaciers of Kangchenjunga.

His main aim, apart from photography, was to cross the Guicha La to the Talung glacier then traverse the saddle on the ridge separating the Talung glacier from the Tongshyong glacier. Once on the Tongshyong, he and his Darjeeling porter would attempt to climb the Zemu Gap above the head of the glacier.

Raeburn had reported on this proposed passage, as it was something he and Crawford had thought to try. They were put off by frequent avalanches and the obvious difficulties of climbing steep ice to the Zemu Gap on the south side. Kellas had been the first to reach the pass from the north and had done so comparatively easily.

If a way could be found to cross the Zemu Gap from the south then it would cut out a lot of portering from Lachen up the Zemu glacier.

Tombazi left Darjeeling in April 1925 with a large party of porters and an Airedale terrier. The party was away for a month. Tombazi was well placed to make such a journey as he was based in India. He was well-connected, being a Fellow of the Royal Geographical Society, member of the Alpine Club (from 1926) and of the Royal Photographic Society in Britain, as well as treasurer of the Photographic Society of India.

The expedition hesitated at the Guicha La due to massive avalanches falling down the south face of the South Peak of Kangchenjunga to the Talung glacier. A camp was set up on the southern edge of the Guicha glacier, where there would be better photographic possibilities. The next morning began in fresh snow and bright sunlight. As Tombazi was preparing his cameras, three of his porters rushed excitedly into his tent – a figure had appeared below. Tombazi saw what he described as a creature, about two to three hundred metres away, exactly like a human being, walking upright, seemingly without clothes, stooping to pull at some dwarf rhododendron bushes before it moved into thick scrub and was gone. Only the footprints were left. The story was taken up by the popular press: it became one of the most famous 'Yeti' sightings recorded. In the final chapter of *Khangchendzonga: Sacred Summit* (2007), the authors write a very interesting and well-balanced assessment.

Tombazi, sadly, had no time to set his camera to get a definitive telephoto image. He suggested this 'wild man' might have been a hermit, possibly a member of some Buddhist community that had renounced the world. In the end he decided it was better to leave the conclusions to ethnological and other experts.

The next stage followed in poor weather and generally bad visibility, with 'ridiculous mists' severely restricting photographic opportunities throughout the expedition. Tombazi claims to have negotiated the Talung and Tongshyong glaciers to ascend the steep icefalls below the Zemu Gap and reach the summit of this pass. There are, regrettably, no photographs to establish whether he had ascended the south side of the Zemu Pass or some other col, as suggested by Jan Kielkowski in *Kangchenjunga Himal* (1999). He seems to suggest (probably after reading Tilman's encounters with the Zemu Gap) that Tombazi's party may have gone over Col 5450 or Col 5422 and that 'It was communicated as an ascent of Zemu Gap.'

Towards the end of his expedition, Tombazi climbed some way up Kabru Dome from the Alukthang glacier to add more images to the splendid set of photographs he had already amassed. On returning home after this 1925 trip, he published privately *Account of a Photographic Expedition to the Southern Glaciers of Kangchenjunga in the Sikkim Himalaya*. The photographs are copies of actual quarter-plate positives neatly pasted into the volume. Many of his black and white prints have been exhibited in the Alpine Club and used to helpful effect in the *Alpine Journal*. So sharp are most of his images that they have helped others plan itineraries on Kangchenjunga and added to our understanding of the mountain's topography and moods.

1926: J.E.H. Boustead (1895-1980)

In May 1926, Boustead claimed to have reached the Zemu Gap from the south, but he might have mistaken his way in poor visibility.

Boustead was born in Sri Lanka, the son of a local tea planter. Educated at Cheam School, he went on to fight with distinction in both world wars. He was an Olympic athlete who had always wanted to take part in an Everest expedition. For that, he needed to have experience of high-altitude mountaineering, hence his visit to the Kangchenjunga area in 1926 and his decision to climb to the Zemu Gap.

He approached via the Guicha La, for which he gave a height of 18,000 feet (5,488m), some 1,800 feet (550m) higher than the actual height of 16,203 feet (4,940m). He descended to the Talung glacier, also in poor visibility, after possibly attaining the Gap. (Boustead's claims were questioned by Tilman: see below.)

1929: Edgar Francis Farmer

In the mid-1920s, Edgar Francis Farmer, from Virginia, finding himself cooped up at work in New York City, decided he was going to climb Kangchenjunga. His preparations were reported in *Mountain* magazine in July 1929 and reprinted in the *American Alpine Journal* (Volume 1, No. 2, 1930). He had, however, never been to the Himalaya, nor had he climbed on ice or snow or any hill above 1,000m.

By April 1929, Farmer was ready to leave Darjeeling, having been helped by G.H. Wood Johnson to gather three experienced Sherpas and seven porters. (Wood Johnson selected Lobsang as sirdar, with Sonam Topgay and Nima Tenduk to organise the journey and purchase food.) Farmer let it be known he was trekking for eighteen days to Jongri, from where he would go up to the Guicha La and the Kang La, mainly for sightseeing and photography rather than mountain-climbing. This was specified as a 'light job'; staff would be paid accordingly.

By 6 May, the party was ready to leave Jongri for Chematang, just below the Guicha La. Farmer went up to the La alone. After several excursions in the area, including a visit to the Kang La and Kabur mountain, the party was resupplied with fresh food from Yuksom and Pemayangtse monastery.

On 16 May, Farmer instructed the sirdar to pack fourteen days of rations and to leave for the Kang La. Two days later they had reached Nepal, to camp near Tseram, where Farmer avoided questions from Lobsang as to whether they had a pass. Farmer urged his team up the Yalung Valley towards Kangchenjunga's south-west face, more or less following the route of

Raeburn and Crawford towards the Talung Saddle. Farmer, well-equipped with crampons, continued, agreeing that the Sherpas should return to the base camp as they had none. Farmer never returned from this excursion. In Darjeeling, the accounts of events were noted by the police from each porter and carefully scrutinised. The investigation exonerated the Sherpas and porters of any blame.

By 24 May 1929 the team had left the moraine covering the lower half of the Yalung glacier and had reached ice. The following events, as told in Lobsang's report to the Indian police, unfolded:

> On 24 May 1929, we were actually on glacier. We saw avalanche falling near our camp. We left at about 7 a.m. and came to our camp at 5 p.m. and halted there. Mr Farmer called this as No. 2 camp. On 25 May 1929, he started at about 7 a.m. and reached the foot of Kanchen Junga mountain at about 4.30 p.m. We made our camp on glacier. It was a difficult place and we made the camping place. We halted there. It was a cold night. In a few minutes our hot tea got frozen. We walked very slowly from Camp No. 1 to this camp which Mr Farmer called Camp No. 3. That night he told us to prepare to leave the camp at 6 a.m. the next morning.
>
> On 26 May 1929, at 6 a.m. we all started. Mr Farmer took coffee, some dry biscuits and ham. He kept no food of any kind in his pockets. He took a small camera, a pair of field glasses, ice axe and two films. He wore three shirts, two coats, three pairs of drawers, three pairs of socks. The nails of his boots having worn out he used a pair of crampons. He took no rope. We walked up together on the mountains but very slowly as we were getting difficulty in breathing. It was 9 a.m. then. I suggested to Mr Farmer that as the sun was up and the snow melting, we may get trouble if we go further and suggested him to return to our camp. Mr Farmer insisted that we must climb the mountain up to 12 noon. We went on. Then we came to a difficult place on snow and ice, sometime down and sometime up. It was like an ice corridor. Then came to a rock and when climbing I slipped and fell down about six or seven feet and injured my back. Mr Farmer told us to wait here for him. He would return at 12 noon after taking some photographs. He gave his Cine Camera to Sonam Topgay to let the spring go when he climbs up. As he started climbing up the mountain Sonam Topgay tried to operate it. We saw

him going up to 5 p.m. As he walked he looked at us several times; and we all called him back but he paid no heed. At 5 p.m. he sat down on the snow. Just then heavy cloud set in and intervened. We waited on the spot until 6 p.m. We thought that he would be returning and we returned to Camp No. 3 and cooked food and waited for him. He did not return. We used his torch light just to show him the camp. On 27 May 1929 at about 7 a.m. we got on the top of an ice hillock and we saw all the way that Mr Farmer went up. We saw Mr Farmer climbing up on the steep snow. This time he was long way up. It was small figure. He was climbing up. The peak of Kanchen Junga was on his left hand side.[1] He got on the top of a mountain when the sun struck the ridge. He crossed that mountain and we never saw him again. Heavy cloud intervened. There were other mountains behind the ridge that he crossed. We waited for the Sahib at No. 3 camp. He did not return. We had no food – all exhausted. There were some dry biscuits of the Sahib which we eat. About 300 feet from our camp we found an old camp for two tents. (Probably the camp of Raeburn and Crawford, who are known to have gone up to Yalung in September 1920.) We thought some Sahibs came here before. There were two empty tins of kerosine oil and also a broken clay pot. There were two heaps of stones (cairns) which we thought were graves.

On 29 May 1929, we waited for the Sahib. As he did not return we left the camp at about 9 a.m. and came to No. 2 camp where we left some food on the up journey. We reached there at about 5 p.m. For want of food we nearly lost our lives. I thought as the Sahib did not return for three days, he must have met his death.

On 29 May 1929, we came to the Base Camp where we met Sonam Chompe and Dam-du and halted there. That night our food was exhausted as we took food for only fourteen days.

On 30 May 1929, Sonam Chompe went to a Gote cowherd and there he exchanged his Chupa Tibetan coat for one Pathi 'Indian Corn'.[2] With this food we came to the hill near Tse-ram and halted there.

---

1 Editor's note, *American Alpine Journal*: 'The description given shows that he went up towards the Talung Saddle (22,130 feet). The party had been working on the Yalung Glacier.'
2 Pathi 'Indian corn': dried cow dung (for making cooking fires).

On 31 May 1929, we camped near Kang La. 1 June we came to Chu-Kar-pang and I sent Sonam Topgay to Jongri and told him to proceed at once to Mr Wood-Johnson and hand over the Cine Camera so that he may be able to develop the films to show the places visited by us and give information about the Sahib.

When we reached Camp No. 3 Mr Farmer had for his food three bundles of pea soup and half a paper box of biscuits, about two spoonful of sugar. We took them as we had no food to eat.

Farmer had been relatively well-equipped but the Sherpas were not. Lieutenant-Colonel H.W. Tobin, knowing the Sherpas well, was able to add more detail about the last sight of Farmer in the *Himalayan Journal* of 1930:

The porters, unable to negotiate steep ice without crampons, went down to prepare an evening meal.

The following day the porters went back up the icefall. They watched again and waited for Francis Farmer. At last, a glimpse of a distant figure moving quite strangely and seemingly, flailing his arms haphazardly. But the lone figure disappeared and Francis Farmer was never seen again.

It is possible that these actions indicate Farmer was now suffering snow-blindness or even cerebral oedema.

Farmer's mother in New Rochelle in the state of New York kept waiting and hoping for her son's return for many years. In her dreams she saw him at the lamasery of Detsenroba in the Yalung Valley living there as a lama, but virtually a prisoner. With all her being she clung to that belief, supported by clairvoyants and spiritualists. When she heard that Professor Dr G.O. Dyhrenfurth was to lead an expedition to Kangchenjunga in 1930, she implored him in letters and telegrams to make a special search for her son:

When the expedition, in compliance with Mrs Farmer's wishes, arrived at the site of Detsenroba, there were nothing but ruins and the graves of the last lamas who had died more than fifty years before. Perhaps Francis Farmer too had reached his nirvana, high on the flanks of Kangchenjunga.

So wrote Dyhrenfurth in *Mountain* magazine. This young man was driven not by fame or fortune but more to explore, fascinated by the unknown.

1935: Reginald Cooke (1901-1996)

A hundred years after Darjeeling was taken over by the British, Reginald Cooke took the train there with his climbing partner, Gustav Schoberth. They arrived with many crates of food and equipment to assist them in climbing Kabru, the prominent mountain on the Singalila Ridge to the south of Kangchenjunga.

Cooke describes his experiences of Darjeeling and the country to its north between the wars. Those drawn to take their furlough from the army, leave from the Indian civil service or from running a business, were fortunate that access was free of fees and most constraints to walking and climbing.

Cooke had once spent two weeks' annual leave in Sikkim; thus began a 'love affair with that enchanted land and its people'.

During the 1920s and 1930s, visitors going beyond the one tolerable road to Gangtok had to apply to the Political Officer there for a permit. These were only granted to a few parties at a time and only to those who could show evidence of physical fitness. Cooke had no objections to that because of the rough going and:

> with a permit, one was privileged to roam fancy free anywhere in this wild and beautiful country. Although it is only one third the size of Wales, besides the peaks, glaciers and towering precipices, Sikkim has dense forested gorges and raging torrents set within great, deep valleys on a gigantic scale.

The only settlement approaching the size of a town before the Second World War was Gangtok, then hardly more than a large village. The main areas of cultivation lay along the rivers of the Tista basin, where terraced slopes produce rice and millet. Nothing had changed much in rural Sikkim since the time of Hooker. Only once, out of a dozen trips into the mountains, did Cooke meet another party of Europeans.

In the late 1920s, Cooke found that the cost of a trek, including food, rest-houses and porterage was only marginally more than the cost of living on the plains.

In 1935, Cooke began seriously to consider climbing a Himalayan peak, such as the unclimbed Pandim (6,691m). This was an obviously difficult peak, much greater than Kolahoi (5,425m) in Kashmir, the only peak he had previously climbed. He was an enthusiastic hiker and a founder member of the Mountain Club, which evolved into the Himalayan Club, of which he became Vice-President after the war.

For the next year, Cooke devoted himself to preparing for his first real expedition to the Himalaya. For trekking in Sikkim, it was generally agreed that October to February were the best months for the clearest views. (Kenneth Mason, in the *Himalayan Journal* of 1936, supported this statistically by summarising meteorological station records from across northern India.) The first half of winter, being the clearest period and therefore a time of low precipitation, enjoyed the extra advantage of offering less chance of avalanche. Cooke had been impressed with Smythe's accounts of such dangers on Kangchenjunga and its satellites, and was determined to avoid them as much as possible, even if the temperature would be lower than during the pre-monsoon. (Not only does the autumn/early winter period mean low temperatures, it also means powerful westerly winds by January/February.) Cooke had often seen an indication of this from Darjeeling, by the long white plumes that stretched from the highest summits. Having decided on October–November–December he began to pay special attention to the design and quality of his clothing and equipment to combat the life-threatening cold and wind at high altitude.

Gustav Schoberth, a like-minded spirit, was a Swiss national working for Siemens in Bombay, and was also a keen mountaineer, having spent several Alpine seasons in Austria and Switzerland. Cooke and he met in Calcutta to discuss plans for Pandim but, at Schoberth's suggestion, changed their objective to an attempt on Kabru. Although it was 600m higher than Pandim, the approach had been well documented by Carl Rubenson, one of the two Norwegians on Kabru in 1907. To Cooke, and no doubt Schoberth, the Norwegians' attempt, with hardly any mountaineering experience and primitive equipment was, as Cooke recounted in *Dust and Snow* (1988), 'one of the finest of its kind in the annals of mountain endeavour'.

Now that they were planning to climb to above 7,250m, Cooke knew from his reading that they could expect temperatures as low as -30 °C and gusts of wind up to 150 kilometres per hour. He put a concentrated effort into designing and making quilted clothes and warm sleeping bags filled with pure eiderdown as well as windproof outer suits.

Knowing how the Norwegians almost starved during their ascent, and how difficult the Everest team had found it to melt snow at the high camps, Cooke exercised much thought in solving these problems. He succeeded in designing, developing and eventually producing a set of nesting pots and pans incorporating a wind-shield, as well as a Primus paraffin stove which had a special burner for high altitudes. It was used exclusively on Everest in 1953 and, two years later, on the first ascent of Kangchenjunga. His system,

'The Cooke's Cooker', could boil water under the most adverse conditions. The Cooke's Cooker, produced at Cooke's Westcliffe Engineering Company, worked perfectly at 12,000m (tested at Farnborough) with air pressure at less than one third of sea level. Cooke thus gave climbers the means to rehydrate and so played a significant role in the success of the first ascents of Everest and of Kangchenjunga.

In mid-October 1935, the two-man Kabru expedition completed its arrangements by employing fifty local men as porters and Sherpas, including a young Tenzing Norgay. Tenzing was noticeable for always smiling and being taller than the average Sherpa. Since he was the youngest, he carried the heaviest load, without complaint.

At Yuksom, their permit, issued by the Sikkim durbar, was checked by the headman. This document vouchsafed his approval and also requested him to assist them through his territory. After spending two days 'bushed' in a valley of dense rhododendron up to six metres high, completely lost in misty weather, they had by 28 October negotiated the Rathong glacier moraines, climbed the Kabru glacier and established Base Camp at 4,787m on the moraine below the Kabru icefall. To the west was the towering east face of Rathong Peak; opposite lay the steep snow slopes leading to The Dome. To the north of their camp lay the main challenge, the 1,500m Kabru icefall, a maze of ice blocks, ice cliffs and crevasses.

Cooke, Schoberth and their five Darjeeling Sherpas, Ang Tharkay (sirdar), Ang Tsering, Ang Thari, Kitan Jigmay and Pasang Phuttar, took a full four weeks to complete a way, with well-stocked camps, through this icefall. Their route followed the east side, under the flanks of the Kabru Dome, which is part of the south-east ridge of Kabru itself.

On 16 November they established Camp 6 at 6,860m at the top of the icefall. They were now below the wide *névé* that feeds the icefall, the same snow that can be seen from Darjeeling, lying between the north and south summits of Kabru.

Schoberth and Cooke set off for the summit the following day, having offered Ang Tsering the chance to join them. He had carried twenty-seven kilos up to Camp 6, and declined as 'I have become a little tired, but it is as the Sahib wishes.' He went down for a rest at Camp 5. Schoberth developed a persistent cough and complained of cold hands, so they decided to take a day's rest. They were cold that night, at -24 °C, despite their 'Everest' sleeping bags.

They recorded their admiration for the Norwegians, who had camped at the same height but in high wind, in temperatures reaching -29 °C, with the difficulty of trying to light a fire of twigs to melt snow and cook food.

On 18 November they set off once more, but Schoberth was again suffering from a cough and cold hands; he descended, having agreed that Cooke should continue.

Cooke reached the wind-blasted, broken, windslab snow on one of the highest ice fields in the world. He turned north, not knowing if the north peak was the higher. He reached the summit, to look over and down 'a gaping abyss to the Talung Glacier below'. Further north, along the main ridge, rather more than two kilometres away, lay another peak, fifty metres higher than Kabru North. (The north peak of Kabru was subsequently found to be twenty-one metres higher than the south peak.)

Cooke's idea, based on his reading, that in November a climbing expedition would be blessed with reasonable weather was borne out by their experience and a comparison with the Norwegian attempt. The Norwegians had reached high on Kabru in October in vicious winds and temperatures down to -32 °C, whereas Cooke was on the mountain a month later with temperatures no lower than -24 °C and no strong winds until the summit ridge.

The expedition retreated over the next few days to Camp 1, where they were surprised to find the ice had been transformed by the strong sun in just over three weeks from a flat surface into *nieves penitentes*, gracefully shaped snow and ice pinnacles a metre high.

At Yuksom, the whole expedition ate a nearly fully-grown pig which the Sherpas bought, killed and cooked over a large wood fire. By the time they had returned to Darjeeling, word of the first ascent of Kabru had reached England, where there was an announcement on the World News and on the front page of the *Daily Telegraph*. Interest in mountaineers and mountaineering was growing.

1937: Cooke and the Hunts

Even before climbing Kabru, Reginald Cooke had thought of exploring Kangchenjunga in order to find a likely ascent route. He had met Captain John Hunt (1910–1998) at Himalayan Club meetings in Calcutta. Hunt followed up their conversation with a letter asking if he could accompany Cooke on his proposed visit to the Zemu glacier side of Kangchenjunga in the autumn of 1937. Cooke readily accepted, having heard that Hunt was an experienced Alpine climber.

Before their departure, Hunt and Cooke had separately visited England and the Alps. By the time Cooke had returned to Bengal, ready for his expedition, Hunt had married Joy (1913–2006), and requested she join the expedition. Despite Cooke's initial reservations, he came to approve of her performance on the expedition. She was not a climber but a considerable tennis player, having competed at Wimbledon. She proved herself fit and strong on the trip, reaching 6,100m with no signs of distress.

The three left Darjeeling on 9 October, arriving at Lachen five days later. They spent a pleasant evening at the Lachen dak bungalow with the Germans Ludwig Schmaderer and Herbert Paidar, and the Swiss Ernst Grob, who were on their way out after a six-week stay on the Zemu glacier. They had been confined to their base camp since 28 September, when two metres of snow had fallen. They had climbed Siniolchu (6,888m) and nearly climbed Nepal Peak but for dangerous windslab snow just below the summit ridge. The newcomers realised they were in for a trying time.

By 18 October they had reached their chosen base camp, just above Green Lake at 4,940m. They sent back all their porters, keeping six high-altitude Sherpas. These included Pasang Kikuli, one of the most famous and experienced Sherpas, and Dawa Thondup, another vastly experienced Sherpa who had been with Hunt in the Karakoram two years before and had only just returned from the ill-fated German Nanga Parbat expedition, of which he was a fortunate survivor. Many first-class Sherpas accompanied this mini-expedition, including Pasang Sherpa, who had recently reached the summit of Chomolhari (7,326m) on 21 May with Freddy Spencer Chapman; Rinzing, a Bhutia, had reached Camp 6 on Everest in 1933 and another Bhutia, Pasang Chakadi, who had been on the 1936 French Baltoro expedition. The base-camp worker was Hawang, a Nepali.

The expedition's scope was limited by the snow conditions which made movement on the glacier laborious and, on the steeper slopes, dangerous. Early on, Rinzing and Chakadi became seriously unwell, with high temperatures. Eventually, Joy Hunt had to escort them back to Lachen. The wind became an impediment, as was Cooke's health. He, on account of Hunt's fitness, felt he had not been given sufficient time to fully acclimatise.

During the month they remained on and around the Zemu glacier they did make several excursions in spite of the conditions. Cooke and Hunt reached the summit of the snow dome on Keilberg (6,091m), just above Green Lake.

On 7 November, Hunt climbed solo to the lower summit of Nepal Peak (7,166m). Cooke had been unwell, so they hoped to return the following day to reach the higher north-eastern summit, but Cooke was weakened by his inability to eat. Cooke was consoled when he and Hunt made the first crossing from the Nepal Gap glacier to the Twins glacier. This was quite a harrowing journey once they had committed themselves by abseiling the icefall above the Twins glacier.

Cooke's party later split from the Hunts and Pasang, who were heading for the Zemu Gap. Hawang was now unwell. Joy, not long back from Lachen, had to escort him to Base Camp while Hunt and Pasang set off for the Gap. It took them three days to complete what was the probable third ascent of the Zemu Gap: the snow on the Upper Zemu glacier had a crust which they would break through up to their waists and where there were shoulder-high hidden boulders.

Once on the col, all difficulties were forgotten as they revelled in the view across the greens and browns of Sikkim to Darjeeling, clearly visible eighty kilometres away. Hunt and Pasang had been surprised to see, at the foot of the final slope below the col, what appeared to be a pair of human tracks in the snow. (In fact, they were standing above the surface of the snow where the compressed snow had been left standing by the action of the wind.) One set of footprints was larger than the other but appeared to be human. They thought at first that the Germans or Grob might have there but Grob subsequently insisted they never went anywhere near the Zemu Gap.

Pasang had originally shown signs of alarm, saying they were Yeti tracks. He recalled that in 1931 similar tracks had been sighted in this area, followed by the death of Schaller and his Sherpa on the north-east spur. Tilman also came across such tracks below the Zemu La in 1938. From these experiences and others, Sir John Hunt wrote in his autobiography, *Life is Meeting* (1978):

> It would seem to be a large ape, possibly an orang-utan, which may have existed until fairly recent times in small numbers at the upper limits of the forests in a few Himalayan valleys. And the memory of these creatures has been perpetuated in stories told to Sherpa children in Sherpa homes.

Between 14 and 19 November, Cooke, Dawa Thondup and Kikuli ascended the Twins glacier towards the north col that lies midway between the Twins, where the north-east spur meets the north ridge. The col lies about 760m above the *névé* of the Twins glacier. The east face leading up to the north col is, for about 600m, steep rock and ice. From a distance it looks totally impassable, but on inspection Cooke and his team, while they did not manage to reach the col, found it less difficult than expected. Cooke thought that given a more thorough attempt they might have found a way up. They made a camp on the face and had reached a point only a few dozen metres below the col. The col was not reached until 1979, from Nepal.

The team made their various ways back to Darjeeling. The Hunts accompanied the main expedition by the walk-in route while Cooke returned via the Simvu Saddle. With him came Dawa Thondup, Kikuli and Pasang with food for four men for twelve days. John Hunt broke trail to the Simvu Pass (5,640m) to see them off.

Their journey was to take them down the Passanram gorge, a feat only accomplished once before by the German Karl Wien and the Austrian Hans Pircher in 1931. Cooke and company, like the Germans, experienced another 'epic': they ran out of food, hacking through the rhododendrons being very time-consuming. Without the good fortune of finding a wild pig (which had died from falling down a cliff), they might never have made it, although the day after their feast on pork they found a track and a Lepcha settlement. They arrived back in Darjeeling on 6 December. Fifteen days after leaving John Hunt on the Simvu col they, and Joy, were reunited: the Hunts were sitting together, writing up their diaries, on the verandah of Telephone House.

1936: Bill Tilman (1898–1977)

This ubiquitous adventurer visited the area twice, first in 1936 during April and May.

Tilman trudged up the Talung Valley on 30 April, with wet snow falling every day. One of Tilman's team was, by 1936, becoming as well-known as many a European: Pasang Kikuli Sherpa. Tilman gives him a special mention in *When Men and Mountains Meet* (1946). Pasang was born in 1911 in Solu Khumbu. He had carried on the three Kangchenjunga expeditions of 1929, 1930 and 1931. He reached Camp 5 on the 1933 Everest expedition and the following year was one of the few survivors of the terrible storm which caused the disaster on Nanga Parbat. Pasang's devotion to duty towards the dying Wieland was honoured by the German Red Cross. In 1939, Kikuli accompanied the American K2 expedition, on which he perished, with conspicuous bravery, trying to save the life of Dudley Wolfe, who had been left stranded at Camp 7 (7,530m). 'He was a great-hearted mountaineer if ever there was one,' concluded Tilman.

Kikuli and two other Sherpas reached the Guicha La, from where they enjoyed their first view of the Zemu Gap and the intermediate snow saddle between the Talung and Tongshyong glaciers. Even at that distance the steep icefall leading to the Gap and a rock wall crowning it aroused misgivings.

They crossed the Talung on a compass bearing in a snowstorm. On the following day, with more snow falling, they camped at 5,515m on rocks to the left of the small glacier descending from Zemu Gap, 300m above them. It was guarded by a steep, intricate icefall, and thereafter by a thirty-metre ice wall. After another camp and a lot of step-cutting, they gave up, due to rotten snow on steep ice and the frailty of the snow bridge over a crevasse. They now had a good view of the final wall two or three hundred metres away up a gentle snow rise:

> It was fully as high as we had feared, all iced and appeared to overhang in places ... If the Sherpas had any regrets at leaving the Zemu Gap uncrossed, they managed to conceal them. Not a moment was lost in packing up ...

They retreated to the Tongshyong to examine the head of the glacier because the Zemu Gap of their experience, wrote Tilman:

> differed so widely from the one about which I had read that I was inclined to believe that a mistake had been made.

But no other break could be found in the east-south-east ridge of Kangchenjunga. It occurred to Tilman that in the misty weather prevailing they had in fact crossed the easy, low col at the head of the Tongshyong glacier which led into the wide bay at the head of the Talung.

They, like Raeburn and Tobin sixteen years before, left the area by the Talung Chu to join the main Tista Valley at Mangen. They too found the dense jungle took them twice as long as expected to negotiate. Tilman, having concluded that the Zemu Gap had never been crossed, was, on reflection, even more enthusiastic about trying again.

1938: Tilman's second visit

Tilman took the opportunity on his way out from the 1938 Everest expedition to devote a couple of weeks to a journey through Sikkim via the Zemu Gap and the Guicha La. From what he had seen, it would need several climbers to reach the Gap from the south but it might be possible to abseil it from the north, with assistance.

With two of his Everest Sherpas and a pony carrying their extra food, Tilman left Tibet to cross the Naku La (5,488m), to camp with Tibetan shepherds at Naku, eight kilometres beyond the pass. He decided not to waste his opportunity: with Renzing Sherpa, he climbed Lachsi (c.6,402m), while the other Sherpa, Lhakpa, remained at a notch below. The ascent took six hours, including a final, very unstable rock ridge covered in snow. The weather was warm: they even experienced rain at 6,100m.

Tilman and his Sherpas then set off south, down the upper Tista Valley. En route to the Zemu glacier, Tilman planned to stay in the government rest house at Tangu on 4 July 1938. He realised, from the number of mules and activity outside the rest house, that it was already occupied – his first response being disgust and the second, curiosity to know who could be in that remote location. On seeing a Nazi flag flying and a large scientific party led by Dr Ernst Schäfer, he was taken completely by surprise. (Tilman had fought on the Western Front only twenty years before. He was later evacuated from Dunkirk.) He remained quite restrained in his comments about the Nazi party members. He mentions Schäfer, a tough-looking, interesting man who went on to 'infiltrate' Tibet as part of the expedition run by Himmler's infamous *Ahnenerbe*, the Ancestral Heritage Organisation (see Chapter 13). Tilman left the Germans to their 'scientific' work that included:

> every breed of scientist known to man: ornithologist, zoologist, entomologist, anthropologist and many other 'ologists' of whom I had never heard.

The next reminder of German interest in the area came on reaching the old German base camp by the Zemu glacier:

> tins and rubbish which would bear comparison with the collections to be found on any of the more popular beauty spots of England, but here it had the shock of unexpectedness.

After camping in the drizzling rain at Green Lake the party followed a cairned track towards the foot of the north-east spur by which the German expeditions made their brave but forlorn attempts to climb Kangchenjunga. Tilman left the track before the spur and took to the glacier, coming down from the Zemu Gap to camp early for a full day crossing of the pass on the morrow.

Everyone was up at 3.30 a.m., brewing tea in the drizzle once again, and then headed into it as it blew down from the Gap. Kellas had appropriately called it 'Cloud Gap'. They were walking on soft snow and took four hours to reach the 5,884m crest. Tilman mentions that they came across a single track of footsteps that also led up to the crest and then disappeared on the rocks on the Simvu side of the Gap. He thought they were no more than two or three days old. These tracks were so obvious that the Sherpas and Tilman wondered if one of Schäfer's party had gone ahead of them. The last time Tilman reported strange tracks was after seeing them on Snow Lake in 1937: it caused, to Tilman's embarrassment, much comment in *The Times* and among scientists and mountaineers. He inquired in Darjeeling on their return: no one knew of any party that had gone anywhere near Kangchenjunga at that time.

The weather on the pass was miserable. Tilman did not spend time looking for the crampons Tombazi said he had left on the north side of the crest in 1925. They now addressed the problem of finding a way down the steep ice wall seen in 1936. To manage that, Tilman had brought a seventy-metre length of alpine line. Even that could not be used on the overhanging ice wall, so they descended towards Simvu. They worked away at lowering their loads from one hacked-out platform to another. Eventually, after jumping five metres over a crevasse, they reached the main Tongshyong glacier at 3 p.m., having taken five and a half hours to descend from the top, and nine and a half hours from their camp on the other side.

With some justification, Tilman questions the veracity of Boustead's crossing in 1926. What a remarkable feat it must have been for his party to leave their camp on the Tongshyong glacier at 3 a.m., cross the Gap, descend to the Zemu glacier and return to camp in time for breakfast at 9 a.m! Clearly, if Tilman, one of the most experienced mountaineers, with Lhakpa and Renzing Sherpa, both 'Tigers', took so much longer than Boustead, it seems that he, and no doubt Tombazi, had crossed a different col.

The group returned to Gangtok and Darjeeling after struggling back up to the Guicha La, then down to drink sheep's milk while sitting by a fire among astonished shepherds camped below the pass:

> We had climbed a mountain and crossed a pass; been wet, cold, hungry, frightened, and withal happy. Why this should be so I cannot explain, and if the reader is as much at a loss and has caught nothing of the intensity of pleasure we felt, then the writer must be at fault. One more Himalayan season was over. It was time to begin thinking of the next. Strenuousness is the immortal path, sloth is the way of death.

1936: Marco Pallis

This expedition to the Zemu glacier area during April and May succeeded in reaching two minor peaks before attempting Simvu. Marco Pallis, the organiser, had a Greek father who had become a British citizen domiciled in Liverpool. Other members of Pallis's team also hailed from Liverpool, including Richard Nicholson, a musician, and the doctor R.C. Roaf. All three shared a strong interest in Tibet and Buddhism and were planning to enter Tibet after climbing Simvu, subject to permission. Yet another Liverpudlian member, J.K. Cook, worked in insurance. He intended to stay on in Sikkim to join Lieutenant J.B. Harrison and visit the Lhonak area along with the sixth member, Freddy Spencer Chapman (1907–1971), a character straight out of a *Boy's Own* annual with his extraordinary powers of endurance and courage, yet a good team player and inspiring leader when the need arose.

The Simvu Five failed to climb the ridge they had set their hopes on as it was sliced through by a huge crevasse or bergschrund six metres wide, ten metres deep and open at each end. With snowstorms blowing through and supplies running out, the expedition dispersed. Chapman and Cooke met Lieutenant J.B. Harrison as the rest of the group departed for Ladakh to engage in Tibetan studies.

Harrison's team then made several attempts to climb Pyramid Peak (7,027m) in June, by its north-east ridge. A fairly substantial subsidiary summit on this ridge was named the Sphinx (6,799m). They climbed it, but after descending its far side for 100m, gave up on continuing to Pyramid Peak. Instead, they climbed Fluted Peak (6,262m) in June, with Harrison, Cooke and Chapman reaching the summit. Harrison thought, since it was so accessible and such an attractive peak, someone (Kellas perhaps?) must have climbed it previously.

This type of expedition was beginning to feature more often, thanks to Kangchenjunga climbers opening up many smaller objectives as they acclimatised. The Germans on the north-east spur climbed several attractive and quite challenging peaks around 6,000–7,000m. Allwein and Brenner climbed Sugarloaf (6,441m) during Bauer's 1931 expedition on 28 August. Hearing or reading of these peaks being climbed encouraged others to try their luck, with members of the Himalayan Club in the forefront of assisting themselves and others to travel and climb in the Kangchenjunga massif.

12

Between the Wars II: the 1930 International Himalayan Expedition

During the decade following the Great War, mountaineers from the German-speaking world had been flexing their muscles on ever more difficult routes, although these had largely to be achieved on a shoestring budget. Climbing in German-speaking parts of Europe consequently became more egalitarian. As numbers increased, so did the severity of the climbs and the ambition to tackle the mountains in the Greater Ranges, including the Himalaya.

The Bavarians were the first to attempt Kangchenjunga, but Günter Dyhrenfurth, in 1927, with his wife, Hettie, inspired by *Round Kangchenjunga*, had also envisaged climbing in the Himalaya. Dyhrenfurth had noted that Freshfield had assessed only two possible ways to climb Kangchenjunga: by the eastern spur or from the north-west, Nepali side. He applied for the eastern spur but his application was bypassed in favour of Paul Bauer's for the same objective. Dyhrenfurth awaited the outcome of the 1929 Bavarian attempt on Kangchenjunga from the east with considerable interest. In 1930, he obtained permission from the Indian government to repeat the Bavarian route. He would, however, have preferred access through untrodden Nepal then up the north-west flanks of the mountain, in order to avoid an ethical clash with Bauer. Dyhrenfurth's expedition would also have to obtain a permit from the government of Nepal, something not granted since Joseph Hooker's expedition in 1848.

Apart from the leader and his wife, the other Swiss members were Marcel Kurz, a cartographer and Europe's most experienced winter climber, and Charles Duvanel, the expedition's cine-cameraman. The three German members were Herman Hoerlin, Helmuth Richter, the expedition doctor, and Ulrich Wieland.[1] Erwin Schneider, a geologist from Austria, completed

---

1   After emigrating to the USA after the Second World War, Hoerlin anglicised the spelling of his first name to 'Herman' (from the original German 'Hermann').

the German-speaking contingent. The expedition included British climbers: Frank Smythe and J.S. Hannah, with George Wood Johnson to help with transport to Base Camp, while Colonel H.W. Tobin was to organise the supply line as far as the Kang La.

It was a very strong team. The geologist, Schneider (1906–1987), had been on the 1928 German–Soviet Alai-Pamir expedition which completed the first ascent of Pik Lenin (7,136m). Schneider often climbed with Hoerlin (1903–1983), a leading young German mountaineer, later elected president of the prestigious *Akademischer Alpenklub Berlin*. By 1927, he had climbed thirty Alpine routes, including several winter first ascents, mainly with Schneider. Frank Smythe had met them during his 1928 Alpine season and recommended them to Dyhrenfurth.

The climbing reputation of Frank Smythe (1900–1949) was enhanced by his prodigious authorship of mountaineering books, one of the first of which was *The Kangchenjunga Adventure* (1930), published within a few months of his return. He became the first fully professional British climber, making a good living from writing and lecturing. (Victor Gollancz, the publisher of *The Kangchenjunga Adventure*, gave Smythe an advance of £1,000, the equivalent today of £70,000, and twenty-five per cent of the book's publication price. *My Father, Frank* (2013) by his son Tony Smythe is highly recommended.)

Ulrich Wieland (1902–1934), from Ulm, had been an enthusiastic mountaineer from his teenage years. The leader, Günter Oskar Dyhrenfurth (1886–1975), was the son of a doctor who had climbed the Jungfrau when sixty-nine. Günter took to climbing from an early age, ascending more than 700 peaks in the Alps and Tatras. He married Hettie Heymann (1892–1972), a well-known international lawn-tennis player of part-Jewish origin; during the rise of the National Socialists, they moved out of Germany and Austria to become Swiss citizens.

All members had experience of winter mountaineering and ski-running as well as summer mountaineering in the Alps. The Bavarians on Bauer's expedition to Kangchenjunga all knew each other before leaving Europe; Dyhrenfurth's expedition members hardly knew each other, except for Schneider and Hoerlin, yet the expedition progressed without personality conflicts. Dyhrenfurth had put his team together to show symbolically that mankind could set nationalism aside to cooperate in such a difficult and challenging project as the International Himalayan Expedition (I.H.E.).

The expedition had received, on arrival in Darjeeling, permission to climb Kangchenjunga only from Sikkim, which meant tackling the north-east

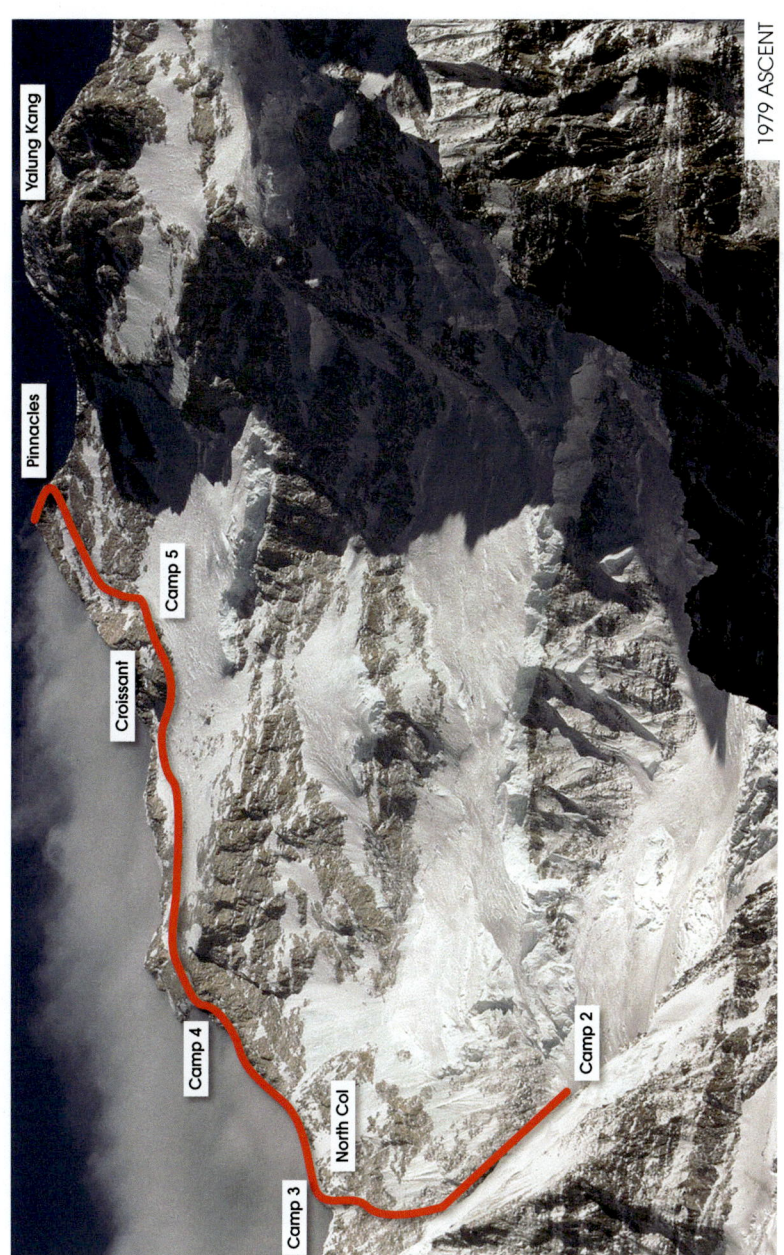

1979 EXPEDITION
Top: Carrying loads.
Above left: Pete Boardman examines his injured ankle.
Above right: Georges Bettembourg.

Top: Joe Tasker sorting out ropes. After a calculation error when planning the climb, the team had to acquire extra rope from a Czech expedition on Jannu.
Above: Doug Scott and Ang Phurba with the camp staff. © *Pete Boardman*.

Top left: The first steep section of the climb.
Top right: Georges on the west face of the North Col.
Above left: Leaving Camp 2.
Above right: Joe climbing in a blizzard.

Top: Doug digging out Camp 3.
Above: Doug at the snow cave.

Top: The route from just below the Croissant (left); the pinnacles are prominent on the right just below the summit.
Above: Georges and Pete after the tent was blown away.

Top: Pete and Joe on the west ridge.
Above left: A brewing storm; the team descended in snow and lightning.
Above right: Approaching the summit.

Top: Doug taking in the view just below the summit.
Above: Pete, Joe and Doug at the summit; they stopped three metres below the peak of the sacred mountain.

spur, as the Bavarians had the year before. After arriving in Bombay and sending off 6,500 kilos of goods by rail to Darjeeling, the Maharajah of Nepal was asked for permission to pass through his country. After arrangements had been made and the I.H.E. was about to set off from Darjeeling, a delightfully sympathetic letter arrived from the Maharajah's Private Secretary:

> Dear Sir,
> I beg to acknowledge the receipt of your letter of the 16th March, 1930, giving information of the formation of an International Expedition to attempt an ascent of and make scientific observations on Kangchenjunga, and requesting permission for the expedition to enter Nepalese territory and approach the said mountain via Kang La, Chumbab La, Tseram, Mirgin La, Kunza and Kangbachen, using the same route on return with the possibility of one party going over to Sikkim by way of Jonsong La. His Highness appreciates your remarks about the international character of the expedition which has for its object the cementation of international friendship and goodwill among the countries concerned, coupled with the augmentation of human aesthetical and scientific knowledge, and desires me to inform you that he gladly accedes to your request. The Nepalese local authorities concerned are being ordered to permit the party the use of the routes mentioned in Nepalese territory.

The Maharajah's interest in the expedition to promote international friendship may have been prompted by his knowledge of Switzerland being in a similar geopolitical situation to Nepal. According to Smythe, Colonel Dawkes, the British Envoy at Kathmandu, greatly facilitated the permission.

The expedition members' joy at pioneering a new route was tempered by having to extend the time to access it. Not only that: it was a difficult journey to cross the Kang La and Mirgin La to reach the Kangchenjunga glacier, whereas the Zemu glacier can be gained in half the time, using good valley paths. The other disadvantage was that winter snow still lay on the Kang La at $c.4,973$m.

Mules could carry for only five days to Yuksom, beyond which the I.H.E. would be entirely dependent on porters to carry its 400 loads. The estimate was for three weeks from Darjeeling to the foot of the mountain at Pangpema pasture. At Base Camp, communication had to be maintained to obtain fresh rations for the retained porters to help ferry loads to higher camps and, eventually, the return to Darjeeling. To ease the situation, the Maharajah was sent a telegram request for permission to purchase porter food at Tseram and Ghunsa. The reply confirmed that local porters would be available to carry supplies.

Colonel Tobin devised a way of avoiding overcrowding at campsites and maximising the number of reliable porters by dividing the expedition into three groups. Wood Johnson and Hannah would manage the first two parties; Tobin would bring up the rear, delivering eighty mule loads to Yoksum, to be collected by 150 porters from the first and second parties and carried over the Kang La.

It was decided to vaccinate all the porters against smallpox. Frau Dyhrenfurth's leading example encouraged the porters to follow. In general, despite the misgivings of some of the team, particularly Smythe, Hettie was a valuable team member, particularly in logistics and communications.

Four sirdars were engaged: Lobsang, Naspati, Gyaljen and Narsang. Lobsang was a Bhotia by birth and, according to Smythe, respected by Sherpas and Tibetans alike and much appreciated by the Europeans for his leadership qualities. Each member had a personal servant. Chettan, the most experienced, was secured by Schneider. Tendechar ('Tencheddar') came highly recommended from 1929 as a superb cook and was duly engaged. Smythe's servant, Nima Tendrup, had reached 7,650m with Sandy Irvine on Everest six years before. In *The Spirit of the Hills* (1935), Smythe extends lavish praise on this Sherpa:

> [he is] endowed with qualities of faithfulness and loyalty; a heart which responds instinctively to the call of comradeship and adventure, that exists in close communion with the noblest aspects of nature, that strives in the cause of others, upwards and onwards, through sweat and weariness, through heat and cold, through hardship and peril. A man to admire, respect and love as a friend.

Crossing the Kang La in deep winter snow tested the expedition: many porters deserted; others fell ill. The three further passes from the Yalung to Ghunsa were similarly difficult for laden porters who had been encouraged to make forced marches through deep snow at high altitude in squalls of snow and hail. A Nepali government official gave assistance from Tseram: he 'encouraged' unwilling porters, and especially the agitators, by threatening them with the Maharajah's displeasure and eventual decapitation!

Smythe quizzed the local yak herder for information regarding Edgar

Francis Farmer, who had bypassed Tseram to avoid attracting attention since he did not have permission to enter Nepal. Smythe was pursuing the request Dyhrenfurth had received from Farmer's mother to search for her son and her suggestion that he could be held prisoner in the 'Detsenroba' monastery. How Mrs Farmer knew of this monastery is not known. Somewhere, the spelling of the monastery had become corrupted. The only monastery in the area is marked on Garwood's 1903 map as 'Dechenrol', but it had been a ruin for at least thirty years. Smythe, Wood Johnson and Nemu, one of the Sherpas who had accompanied Farmer, visited the ruins near Tseram, but found no signs of the American.

At Ghunsa, the government representative arranged for vital porter food to be available along the final journey to Base Camp through Kambachen and Lhonak to Pangpema. Base Camp was established on 26 April, though later than planned for sufficient time to acclimatise and climb the mountain before the monsoon snows blew in. Unlike the Bavarians, who had chosen to walk in during the monsoon and climb in the autumn, Dyhrenfurth had taken note of the British experience of three visits to Everest and opted for a much earlier walk-in during the pre-monsoon period.

From Pangpema, at 5,141m, with supplies still trickling in from below, the spearhead climbers set off up the Kangchenjunga glacier to find a possible route towards the north ridge and the main summit. This route was prompted by Freshfield in *Round Kangchenjunga*. He recommended that a search to the left, towards the saddle that connects Kangchenjunga and the Twins might be worth considering but was doubtful that 'any coolies will be found to carry ... above the lowest icefall'.

By 1930, climbing on the highest Himalayan peaks was led by the belief that only the progressive siege of a mountain, with porters bringing up supplies to stock camps, was considered practicable. The steep, west-facing ground leading directly to the North Col of the north ridge was therefore hardly considered, despite it being free from avalanche danger. The only possibility was to tackle the ice cliffs and snow terraces to the right and hope they would be free from avalanche during the climbing. Once the lower terrace had been reached the intention was to establish Camp 3 from where, by traversing left, the north ridge could be gained.

For eight days, Smythe and Wieland, then Hoerlin and Schneider lay siege to the ice wall leading to the lower terrace on the north face. On 8 May, Hoerlin, now with Dyhrenfurth, reached the crest of the 220m cliff, hammered in ice pegs and fixed ropes to them. (These pegs had been designed by the great German mountaineer Willo Welzenbach. They consisted of a barbed, flat metal strip, between twenty-three and thirty-one centimetres long, with an iron ring at the top end. They were hammered into the ice and a rope clipped into the ring with a karabiner. Depending on the quality of the ice they offered great security for belays and fixed ropes. The 1929 and 1930 expeditions were first to bring such artificial aids to the Himalaya.)

The party then descended, along with Kitar and Chettan, who had been close behind Dyhrenfurth and Hoerlin, while preparing the ice to improve the steps for future porter carries. As they were descending to Camp 2, and pleased with their efforts, Dyhrenfurth recalls that:

> Chettan stopped me, pointed to the cliff and said, 'Sahib, no good.' Then he tried, mainly by signs, to indicate to me how dangerous he thought the sector.

Dyhrenfurth, with respect and without ambivalence, writes that:

> he was one of the best climber-porters and, at the time, far ahead of us in Himalayan experience ... everybody was very happy in Camp 2 that evening, all expectation that the way was now open for much easier climbing towards and onto the North Ridge.

Following a warm night, morning dawned grey and misty with snow flurries; Dyhrenfurth and Chettan set off first. They were carrying pitons and ropes to put the finishing touches to the porters' route. Just before the cliff, a giant crevasse had to be crossed by a narrow snow bridge protected by ropes. Dyhrenfurth paused after crossing the crevasse to inspect their route above, only to hear a cracking sound and then to see from the top of the cliff to his right:

> an ice wall perhaps one thousand feet wide ... toppling forward quite slowly ... before the huge face broke and came crashing down in a gigantic avalanche of ice. The impact of the ice fragments whirled up a dust curtain of snow and ice, which with incredible rapidity broadened out into a solid, perpendicular wall.

With little hope of escaping the blast, he ran to his left through deep powder snow, shielding his face with his arm against the dust cloud. The uproar around him was fearsome; he lay in the snow awaiting death:

> Suddenly there was an uncanny silence ... My subconscious mind began to marvel at my still being alive.

Dyhrenfurth had survived, as had the other five members and twelve porters who were either still in Camp 2 or on their way up to the ice cliff. Some had been caught only in the peripheral cloud of powder snow and were able to free themselves easily. Chettan did not survive. His rucksack was heavier than Dyhrenfurth's and so he was some way behind. Although directly in the line of ice blocks that had missed Dyhrenfurth by five metres, he was quickly located and dug out. He was not breathing but showed no sign of external injury. At the request of his Sherpa friends, he was laid to rest in the glacier. Dyhrenfurth concluded:

> All honour to the memory of Chettan, one of the best of the 'Tigers' – quiet, reliable and full of courage.

A growing feeling that this area of the mountain should be abandoned was led by Smythe. Dyhrenfurth wanted to remain to try another way; he was able to persuade the rest of the party to do so. On one hand, Dyhrenfurth was idealistic in his strategic sense, but in the field he remained somewhat controlling: first thing in the morning, he would wake the camp with a bugle call; when the bugle was stolen he produced another one!

Under orders, the party next attempted the north-west spur above the western arm of the Kangchenjunga glacier. On closer inspection, this looked hopeless as a summit route. Every climber took turns in attempting the north-west spur, with Schneider and Wieland reaching the top of the first big tower at 6,400m. The climbing had, in Smythe's opinion, been harder than anything ever done at that altitude and was far too difficult for porters. The rock was rotten and very exposed to the weather. The expedition left Kangchenjunga for secondary objectives, wiser for having experienced avalanches crashing down the flanks of Kangchenjunga and the surrounding peaks every day: Kangchenjunga was not, they decided, an ordinary mountain.

Before leaving their glacier camp, Smythe and Schneider, thanks to good acclimatisation, made a mostly ski ascent of Ramthang Peak (6,601m). (This had been labelled 'Madonna Peak' on account of its beautiful, shapely summit rising above 6,000m. Dyhrenfurth, following the practice of the Royal Geographical Society, rechristened it Ramthang Peak since the snow and ice on its south flanks feed the Ramthang glacier.)

Schneider and Wieland explored a route on the Nepal side of the Nepal Gap. They solved the problem of crossing the range here by ascending a subsidiary of the Kangchenjunga glacier which falls from the ridge between the Nepal Gap and the 7,155m peak to its north. They camped just below the ridge, which they easily reached the following morning. On the other side, to the north-west of the Nepal Gap, they discovered a short, steep snow slope leading to the Zemu glacier. They had, after many attempts by Kellas, found a way of crossing the main Singalila Ridge, not exactly over the Nepal Gap itself but a few hundred metres to the north. As Wieland was feeling unwell, the tireless Schneider set off on his own up the ridge, sometimes on a very steep rib of snow and ice in perfect condition for him to solo to the summit. It was later named by the expedition 'Nepal Peak'.

On 30 May, Schneider, Hoerlin, Wood Johnson and Smythe set out for Jonsong Peak (7,462m). Owing to Wood Johnson becoming ill, neither he nor Smythe were able to reach the summit. This was the highest peak climbed by that date. A few days later Smythe did reach the summit, along with Dyhrenfurth, Wieland and Kurz.

Hoerlin and Schneider then moved north to climb the more difficult Dodang Nyima (6,920m), at the head of the North Lhonak glacier, before leaving for Darjeeling. This remarkable pair then rejoined the expedition on its walk-out to Darjeeling via the Jonsong La and Gangtok.

The International Himalayan Expedition was widely reported in newspapers, books and journals. In *The Kangchenjunga Adventure*, Smythe stressed the main problem was not extended supply lines or ferocious storms but the continuous nervous strain of never feeling safe from avalanches. He said the time spent on Kangchenjunga was the most nerve-racking he had ever experienced on any mountain.

13

Between the Wars III: the German Expeditions

1929: the north-east ridge of Kangchenjunga

The north-east ridge was first identified by Douglas Freshfield in 1899; he described it as:

> a marvel of mountain architecture: it springs from a low mass or pedestal of splintered granite and flies up as a mountain arête, of a length and steepness which defy Alpine comparison ...

Thirty years later, a strong, well-equipped team of climbers had settled in at their Green Lake base camp. They were the first cohesive group of hand-picked, experienced mountaineers, with the latest clothing and equipment, to visit Kangchenjunga. They were strongly motivated to climb for Germany.

A founder member of the Himalayan Club, Kenneth Mason, recalls that the Club was contacted in 1928 by Rickmer Rickmers, the veteran mountaineer from Heligoland, requesting advice for a young Bavarian climber, Paul Bauer (1896–1990), who was planning to bring a group of mountaineers to the Himalaya. 'They want to test themselves against something difficult,' wrote Rickmers, 'some mountain that will call out everything they have got in them of courage, perseverance and endurance.'

Paul Bauer wrote in *Himalayan Quest* (1938) of the time just after the Great War, a period of desolation back in 'civvy street' when he began to find contentment in the mountains:

> They had the power to restore that which the town environment threatened to steal.

Bauer declared that the German expeditions to Kangchenjunga and Nanga Parbat were strongly motivated to prove that Germans could rise above their humiliation following defeat and the Treaty of Versailles.

Eventually, nine mountaineers set out for India: Eugen Allwein, a physician from Munich who had made the first ascent of Pik Lenin (7,122m) the year before; Peter Aufschnaiter, an agricultural scientist from Kitzbühel who went on to be immortalised in Heinrich Harrer's *Seven Years in Tibet* (1953); the photographer, Julius Brenner, a chemist from Kaufbeuren; the 'head chef', Dr Ernst Beigel, who took part with Bauer in the 1928 Pik Lenin expedition; Wilhelm Fendt and Karl von Kraus from Munich; Joachim Leupold from Arnstadt, who had made the first longitudinal winter crossing of Mont Blanc; and Alexander Thoenes, an engineer from Speyer. All the team, including Bauer, the expedition 'Führer', had made many significant climbs over the previous years. Bauer was very proud of them. He wrote:

> They were a band pledged together for life or death, proud, determined, self-confident men, united in fanatical devotion.

The 1929 German expedition arrived in Calcutta at the end of July, to be met by Mason and other members of the Himalayan Club.

'By a curious coincidence,' wrote Mason, 'Bauer and I had fought each other in the trenches 100 yards apart in France in 1915. The mountains have drawn us together since and we remain good friends.'

Colonel H.W. Tobin and Edward Oswald Shebbeare had arranged the transport for the expedition, as far as the Zemu glacier. Thanks to vastly improved communications in Sikkim and the organisation provided by Tobin and Shebbeare, who were able to gather many more experienced porters than in Freshfield's time, the Germans were settled at their base camp less than three weeks after their arrival in India, although they had been in the unenviable position of not knowing, when they left home, which mountain would be made available to them by the Indian government. In Calcutta, however, the authorities whisked them through customs without charging duties on their equipment.

From their reading, the Germans knew about the previous encounters with the mountain and were especially familiar with Douglas Freshfield's observations. They had little choice of route, since they were allowed access only from Sikkim. The Talung glacier side of Kangchenjunga was ruled out because of obvious avalanche dangers. The north-east route was similarly impracticable from the Nepal Gap, as it would have made for too long a haul to surmount the Twins to reach the north ridge. They therefore set off from Green Lake (which remains 'Green' as it is fed in part by geothermal activity), at 4,360m on 18 August, to carry out their own exploration. Bauer's time spent on the Western Front had obviously made a strong impression upon him; war imagery permeates his description:

> Six men in solar topees with ice axes under their arm – it looked very much like an advance, as if the infantry were moving up to attack and however much war may be solemnly denounced, yet my old fighting spirit rejoiced.

The group was split, with one party intending to climb Simvu via the Simvu Saddle: they were driven back by bad weather. The second group climbed to the base of Kangchenjunga itself to reconnoitre the main east ridge which continues east to Simvu and Siniolchu. The lowest point of this great ridge is the Zemu Gap. It was dismissed as impracticable because of its sheer length and vulnerability to westerly winds. The party did appraise the North Col as a means of reaching the north ridge but abandoned that route as it seemed to involve 1,000m of steep rock, snow flutings and ice. Going all the way from the Nepal Gap was considered but dismissed after recognising the huge logistical problems involved. The only remaining option was the spur which rose out of the Zemu glacier to join the main north ridge about one and half kilometres horizontally and some 500m below the summit. (The adoption of the label 'north-east' has differentiated this spur from the main east ridge of Kangchenjunga.)

Eventually, the team established an Advance Base Camp on the Upper Zemu glacier to begin their climb on the south-facing side of the north-east ridge. They were confronted with a very steep wall of rock and gullies choked with ice before they could attain the crest of the ridge at about 6,000m. After that, the ridge reared up in a series of snow and ice-encrusted rock towers and steps. Above about 6,850m, the going became easier, across a broad whaleback.

The climbing up the wall began on 26 August, from Camp 6 at 4,950m. During the next three weeks, some of the most difficult climbing that had ever taken place in the Himalaya was performed on the south side of the north-east spur, just to reach the crest of the ridge. On the ridge, huge problems, which seemed so inconspicuous from below, had to be negotiated. The route necessarily chosen was partly protected from wind until they arrived on the crest. (The ice and snow on Kangchenjunga melt to form ice that is aerated and resembles a fretwork of lace upon which it is impossible to belay safely.)

After taking time out for bad weather and alternating climbing pairs, the ridge crest was reached, but only after an enormous amount of work to make the route passable for porters and allow a retreat in bad weather. More or less a staircase was cut through the snow and ice, with a tunnel forced through a cornice of ice blocking the way just below the crest. At high altitude, this work was found to be very tiring, so the leader was rotated every fifteen minutes. Allwein describes this:

> There was nothing for it but to force a shaft vertically upwards from the niche ... it was fearfully heavy and exhausting labour for one had to wedge one's self inch by inch into the shaft as it took shape; as the man at work hacked the snow away, it fell on his face and shoulders and no matter how closely fastened his clothing the melting water found a way down into it.

The following day, after improving the route by hacking off masses of crumbling snow and ice:

> We set to work again on the tunnel, which Beigel managed to complete at last after another hour's work. He came upon another exiguous ledge under more overhanging snow but had to come back after trying to work his way along it to the left. After that Thoenes hacked a way up towards the final cornice in a snow groove, which in the end I succeeded in surmounting ... by then the working day was running out.

On 3 October, from an ice-cave at 7,000m, Allwein and Kraus left to reconnoitre the now much-easier-angled section. They reached 7,400m. The expedition was apparently now well-positioned, with ice-caves to accommodate at least half a dozen climbers and Sherpas at camps 8, 9 and 10.

On the following morning, with most of the expedition members at Camp 10 and full of summit expectations, it began to snow heavily; the thrust was temporarily halted. Kraus, Thoenes, Chettan and Lewa descended to save food and fuel and to assist with the logistics lower down the route, where Brenner and Fendt were moving supplies uphill.

The snow continued to fall for two days. Beigel and Aufschnaiter descended to ensure four days' food and fuel remained available for Bauer,

Allwein, Ketar and Pasang, who would try to reopen the route beyond Camp 10. As it turned out, the snow was just too deep and difficult: heavily laden climbers sank through a hard crust into soft snow up to their waists. In two hours, they climbed only seventy-six (vertical) metres on a fairly gentle slope. They continued without packs, just to see more of the surroundings which:

> looked so wild and chaotic from the gorges below [but] were now spread out peacefully in an almost orderly fashion below them ... Simvu, Siniolchu and the Twins lay below us. They were beautiful to look at ... all the danger they represented had disappeared.

They were soon to be enveloped by danger. At Camp 10, the atmospheric conditions became:

> eerie, even sinister with great banks of cloud massed about 30,000 feet high in the south and with the skies above the Zemu Gap and Simvu strangely sea green.

A massive amount of snow now fell on the German team, strung out along their climb; the entire expedition was placed in great danger. It took two weeks for them to descend safely through all the camps to the Zemu glacier, and to wade through the fresh snow to Base Camp. This journey was made more difficult by Beigel, with severely frostbitten feet, having to be carried on an improvised stretcher. The mountain and Base Camp were cleared; the party arrived back in Lachen on 20 October, in pouring rain. This epic retreat only succeeded because of the group's enormous experience, mountaineering skills and ability to work well as a team. They had pushed their resources of manpower and materials to the limit, without going beyond it. The editor of the *Alpine Journal* summed up their achievement:

> It was a feat without parallel perhaps in all the annals of mountaineering.

Although unsuccessful, no climbing of such difficulty had ever been achieved before in the Himalaya.

1931: Bauer's second expedition to the north-east spur of Kangchenjunga

Bauer came to see Kangchenjunga as his and Germany's *Schicksalsberg* (mountain of destiny). In 1931, he increased his team to eleven climbers, including Allwein, Aufschnaiter, Beigel, Brenner, Fendt and Leupold from 1929, as well as four additional younger members: Hans Hartmann, Hans Pircher, Hermann Schaller and Dr Karl Wien. The attempt was still to be made in the pre-monsoon and monsoon periods.

Bauer had been quite critical of Dyhrenfurth's organisation and his Jewish wife – and even of Hoerlin and Schneider for placing only their Swabian and Tyrolean flags on the summit of Jonsong Peak. Bauer, leaning ever closer to the National Socialist hierarchy in Germany, was beginning to influence German and Austrian climbing politically. He began to exclude climbers of Jewish origin from the National Alpine Club.

Bauer's second expedition to the north-east spur arrived in Darjeeling later than planned; further delays were caused by porter unrest. Just as the problem was being settled, Lobsang, the sirdar of the 1924 Everest expedition, and the porter, Babu Lau, both died of a mysterious tropical fever. Leupold and Fendt remained ill and ineffective throughout the expedition.

Nevertheless, Camp 8 was reached after climbing valiantly through storms, wet snow and crumbling rock. The temperatures were much higher than in 1929. On 9 August, Sherpa Pasang fell down a gully below Camp 8, pulling off his rope-mate, Hermann Schaller, held by a second Sherpa, Tsin Norbu, until the rope snapped. Pasang and Schaller then fell nearly 600m to their deaths. Bauer himself strained his heart on 16 September but with the drug cardiazol he managed to continue to function. Allwein frequently experienced excruciating sciatica. The Sherpas' morale was beginning to fall. Despite these setbacks, the climb continued until, on 17 September, Hartmann and Wien reached the summit of the north-east spur (7,702m), only to realise they could go no further. Although the ground looked reasonably gentle, the spur dipped some sixty metres before it rose to join the main north ridge in a 150m concave slope which was dangerously avalanche-prone.

'It was,' wrote Hartmann, 'in a truly shocking state ... about one and a half feet of powdery snow lay lightly on the hard frozen strata ... there was no prospect now of any improvement in the slope as it snowed daily for several hours.'

There was no way of avoiding it. After two days' inspection by Hartmann and Wien, then Pircher and Fendt, the expedition went into retreat. With supplies almost exhausted, there was no time to sit it out. They withdrew in good order. They had climbed harder than anyone before them on a high Himalayan peak.

Charles Evans later summed up this superhuman effort by Paul Bauer and his team as 'a classic example of skill, courage, energy and judgement'. For forty days, at a height of over 6,000m, they had carried heavy rucksacks and undergone severe step-cutting with hardly a day's rest. Despite this, they made the most of the return to Darjeeling by taking different routes: Leupold and Aufschnaiter, with their porter, trekked from Base Camp up the Green Lake glacier to its head then over the Podon La, before walking to the Tibetan frontier then traversing east to the Donkia La, returning to Darjeeling via the Lachen Valley. Allwein and Pircher, with their porters, made the first crossing of the Simvu Saddle before descending south-east to the Passanram glacier, from where they enjoyed excellent views of the south-west face of Siniolchu and the Rock Needle. From there, they struggled down gorges for seven days, to cover just sixteen kilometres as the crow flies, before eventually reaching the Talung Valley. (They emerged where Pangong and Sanven are indicated on the map, although these places do not actually exist.) There was only jungle. Just before Sakyang, on coming across a path, they met Lepchas armed with bows and arrows. Finally, after continuing down the Talung Chu, they reached Mangen, on the Tista river. Here, they rejoined the main body of the expedition that had slowly marched from the Zemu Valley, in all its autumn colours.

The next foreign visitors to climb the Kangchenjunga massif were Reginald Cooke and friends (see Chapter 11).

1936: Bauer's third expedition: the background
Thanks to the stimulus of National Socialism, German mountaineers remained at the forefront of Himalayan climbing during the 1930s. Paul Bauer had sensibly chosen like-minded climbers who shared his love of climbing and devotion to their Fatherland.

Their country was changing fast. When they left in 1931, members were hearing of the worldwide economic depression that had hit Germany particularly hard. By February 1933, Hitler had established absolute political authority. His domestic aim was to enforce *Gleichschaltung*, the ideological control over all aspects of civil life, including arts and sports.

The universally well-regarded German and Austrian Alpine Club, with nearly a quarter of a million members, was an obvious Nazi target. Unlike the British club, the German and Austrian club welcomed men and women, regardless of class, politics or climbing achievements.[1] The *Alpenverein* was, from the first, committed to preserving the mountains from commercialisation. From late in the nineteenth century, antisemitism, especially in Austria, led to the prohibition of Jewish applicants. In 1921, Jewish mountaineers formed a new section in Vienna called *Donauland* ('Land of the Danube'). In 1924, at the annual general *Alpenverein* meeting, eighty-nine per cent of the representatives voted to expel *Donauland* rather than upset the Austrians. According to Bettina Hoerlin (Herman's daughter), the 1924 decision is considered today to be the darkest chapter in the history of the *Alpenverein*.

On Hoerlin's return from South America, he took up the invitation to join the Executive Committee of the Club at this most difficult time. Hitler had appointed Hans von Tschammer und Osten as *Reichssportführer* (Reich Sport Leader), with orders immediately to apply *Gleichschaltung*. Committed climbers, however, were unwilling to take orders from anyone. In the *Alpenverein*, sufficient mountaineers remained to continue to put the brotherhood of the rope above the authority of the state.

The Sport Minister, aware that he was not dealing with the usual compliant representatives of competitive sports, agreed not to dissolve the club if it adopted National Socialist doctrine with regard to Jews. The price of independence for the world's largest mountaineering club was to exclude Jews and other 'non-Aryans'.

Hoerlin, and a caucus of like-minded club members, tried to resist Nazification. Paul Bauer favoured appeasement of the National Socialists. As a reward for his support, the Sport Minister increased Bauer's authority by creating the German Mountaineering and Hiking Association and putting him in charge. Bauer continued to promote Himalayan climbing as the way to rebuild the morale and strength of the German nation, by establishing the state-supported 'Reich League of German Mountaineers' to:

> lead German mountaineers to a consciousness of their high calling,
> to guide the rising generation, that they may learn to be fearless.

The new organisation's vision, Bauer informed a meeting of the *Alpenverein*, was based on:

> the heroic world view for which our Führer Adolf Hitler has fought,

---

1   The German and the Austrian societies had merged in 1873 to form the German and Austrian Alpine Club (DÖAV).

and throws our mountaineering activities into the proper light before the eyes of the public.

Bauer's objectives were to absorb the 240,000 club members into the Nazi state structure, rid all sports clubs of Jews and to control all future expeditions leaving Germany. According to Bettina Hoerlin, her father 'vehemently opposed each of these objectives'. Hoerlin and Bauer's mutual animosity dated back to 1928, when:

> Bauer off-handedly rejected Hoerlin's enquiry about joining his 1929 Kangchenjunga expedition. Hoerlin, with a reputation as an independent thinker, was not welcomed by the group of climbers cultivated by Bauer.

This politicisation of German mountaineering detracted from the former goal of keeping the Alpine environment free of commercialisation. Hoerlin, in particular, found that frustrating and demoralising, as he wrote to his wife Kate:

> I am no longer capable of participating as I had beforehand. One group sits there with their brown Socialist Army uniforms … others in black SS uniforms. Many are totally pig-headed. What do I want with them?

With Bauer threatening real trouble for Hoerlin, he resigned from the Club's Executive Committee in 1937.

Bauer played an important part in the exploration of the Kangchenjunga massif. He was also a climber of great determination and courage and gave other younger climbers their first opportunity to climb in the Himalaya. He cannot be ignored, nor his part in the Kangchenjunga story disregarded because of his strong support for the Nazis. He was a product of his time: someone who had come out of the Great War totally disorientated. His authorship was well-regarded: he won the gold medal in the Arts competition of the 1932 Olympic Games in Los Angeles for his *Am Kangchenzonga*, an account of his second expedition.

After 1931, Bauer gave up on Kangchenjunga, partly because he found it difficult to raise funds for a third attempt but also because Nanga Parbat came to be Germany's preferred 'mountain of destiny'.

1936: Bauer's expedition to the Kangchenjunga massif

During the monsoon, Bauer took a small group of friends to Sikkim, ostensibly on a training exercise for an expedition to Nanga Parbat the following year. Now forty years old, he tried a return to basics after asking himself:

> Are numerous mountaineers, Sherpas and porters really necessary or would it not be more rewarding to climb in the Himalaya as a party of three or four amateurs and a consequent reduction in the number of porters?

As a result of Bauer's new thinking, his only companions were Dr Karl Wien, in charge of scientific and photographic studies; Dr Günther Hepp, the doctor; and Adolf Göttner.

They left Germany on 10 July, pitching up at their Zemu Base Camp on 16 August. Bauer wrote that:

> it was good to be back in Sikkim, full of pleasant memories to us; everywhere we met old acquaintances who again helped us joyfully.

Paths to Lachen were found to be much improved, with little risk to the mules from landslide. Lachen had developed into a village where porters could be hired on the spot, making for greater ease in supplying the expedition with fresh food.

Six Darjeeling porters accompanied the German party: Hishey Bhutia, Tewang, Nima Tsering II, Mingma Tsering, Mingma Bhutia and Phurba Bhutia. They would help the Germans break trail through the fresh snow and carry loads to those peaks which had been largely ignored when they were besieging the north-east spur of Kangchenjunga in 1929 and 1931. Knowing of these attractive satellite peaks inspired this lightweight expedition to attempt them in 1936. The main objective was Siniolchu, the icy peak rising like a gothic cathedral well to the east of Kangchenjunga.

They first explored new ground south-east of Siniolchu, at the head of the Zumtu glacier, which enabled them to reconnoitre the easiest way up. During early September they made an unsuccessful attempt on the eastern summit of the Twins (7,350m) and Tent Peak (7,365m) by its south-west ridge. They had to give up on account of avalanche danger but did pass over the summit of Nepal Peak (7,177m) to reach Tent Peak's south-west ridge.

With the onset of better weather, the climbers, plus Nima Tsering and

Mingma Tsering, left their base camp on 19 September for the north-west ridge of Siniolchu (6,888m). By 21 September they had established an advance camp at the base of Little Siniolchu (5,701m), where they left the Sherpas. The following day, the climbers bivouacked at 6,200m in their Zardsky bivouac bags on the ridge between Siniolchu and Little Siniolchu.

On 23 September they set off up the north-west ridge. Bauer and Hepp decided to drop out but remain in support of the two now unladen men going for the summit. Göttner and Wien reached the summit by 2 p.m. They returned to their companions to spend another night on the ridge in bivouac bags, with their feet stuck in their rucksacks.

They had returned to Base Camp by 25 September, having made the first ascent of one of the most beautiful mountains in the world and, as befits such a peak, in Alpine style without all the clutter of fixed ropes between static camps.

After a suitable rest, Göttner and Hepp climbed Simvu North Peak (6,547m) on 2 October, completing what the British had failed to do the year before. After a further spell of poor weather, Göttner and Bauer climbed several peaks around 6,100m, including Podon Peak, Green Lake Peak, Black Peak and White Peak. Meanwhile, Wien had crossed the Simvu Saddle, hoping to extend his 1931 survey of the Zemu glacier to the south of Siniolchu. Unfortunately, his photo-theodolite was damaged in an accident.

Bauer had achieved his main goal of passing on his Himalayan experiences to younger German climbers, all three of whom formed the core of the 1937 expedition to Nanga Parbat, where all three perished.

1937-1939: the German-Swiss expeditions
Doom and gloom were not universal in the climbing world during 1937. Cooke and Hunt enjoyed a useful visit to the Zemu glacier – particularly Cooke's exploration of the Upper Twins glacier and the steep wall of rock and ice below the North Col of Kangchenjunga (see Chapter 11).

Pasang and Freddy Spencer Chapman's 1937 ascent of Chomolhari demonstrated how fast climbers can reach well above 7,000m with little in the way of equipment and team support. What Chapman achieved could be replicated on Kangchenjunga, less than 120 kilometres away. On Kangchenjunga, more than elsewhere, climbing needs to be fast and efficient to reduce the time one is exposed to avalanches. Reading Chapman's account was an inspiration to those few high-altitude mountaineers of that time.

The mountains of Sikkim, and the Sikkimese themselves, seem to have held a magical power over certain individuals who simply could not keep away. Three such people included the Germans Herbert Paidar and Ludwig (Wiggerl) Schmaderer and their friend Dr Ernst Grob. Their German–Swiss expedition spent six weeks in the Zemu glacier area during August and September of 1937, with the principal aim of finding a climbable approach to Kangchenjunga. They attempted to make the first ascent of Tent Peak (7,365m) by the south-west ridge but failed owing to dangerous monsoon conditions. The same soft windslab snow also prevented them climbing the Twins (7,150m), although they did make the second ascent of Siniolchu on 26 September.

In 1939, the three friends launched their second German–Swiss expedition to Sikkim, sensibly going during the pre-monsoon season. Their main objective this time was to climb Tent Peak (7,365m). This magnificent northern outlier of Kangchenjunga had remained unclimbed even after valiant attempts by Göttner and Wien, their own attempt the following year as described above, and Sir John Hunt and Reginald Cooke's attempt later in 1937.

On their first attempt in 1939, their 'orderly' (porter) Kandova fell just above Camp 2 on 20 May and broke his leg, forcing them back to Base Camp. By 23 May they had seen to the evacuation of the injured porter and were ready to make another attempt. They knew the only way for them was to reach Nepal Gap (6,302m), then traverse Nepal Peak (7,177m).

On 26 May they put up Camp 5 at 7,012m, only three quarters of an hour from the summit of Nepal Peak. Here they left Ila, an Everest 'Tiger', with the porters. Better weather arrived on 27 May: they reached the small summit cone of Nepal Peak, from where they could see all the irregularities to negotiate along the ridge to the summit of Tent Peak. (Nepal Peak was first climbed solo by Erwin Schneider during the International Himalayan Expedition of 1930.) The two returned to their 'orderlies' at Camp 5 to prepare for the push to the summit of Tent Peak.

On 28 May they set off with rations for three days and a lot of climbing equipment to tackle the snowed-up rock towers along their route. They instructed the support staff to wait at least five days for their return. (This plan later led to some criticism from Himalayan Club members concerned with porter welfare.)

They spent a hard day's climbing in magnificent weather. Camp 6 was placed on a saddle after finding themselves committed all day above 7,000m, traversing a narrow snow-crest in bright sun with some of the finest mountains in the world extending above the billowing cloud inversion.

On 29 May, they progressed from their highest camp along the ridge, up steep pillars of granite, to finally reach the summit of Tent Peak. After a few minutes identifying all the peaks they had climbed or attempted, and others beyond, as far as Everest 120 kilometres to the west, their thoughts turned to home and their families.

They spent a cold night in Camp 6, then took four long, tiring days to descend to Base Camp, having climbed their much-attempted peak in a single push with help from their Sikkimese 'orderlies', Ang Tsering (Pansy) the Sirdar, Ila, Arjeeba, Ila (sometimes 'Ilia') Tensing, Ang Karma and Gendin Bhutia.

This *Drei im Himalaja* ('Three in the Himalaya') team went on to make the first crossing of the Langpo La and the first ascent of the southern summit of Langpo Peak.

Grob, who wrote the account of the expedition for the *Alpine Journal*, wondered if, on his return to Sikkim, it would be with ice axe or, most likely, a botany box to:

> observe men, beasts and plants in wonderful Sikkim, with perhaps two orderlies – how splendid! And then, through the primeval forests ...

That opportunity would not arise for some time. On arrival in Gangtok the expedition members were informed that on 3 September, war had been declared. Grob, being a Swiss national, was given the required paperwork to return home. Paidar and Schmaderer were interned.

The following year, the book of their expedition was published in Munich by Bruckmann. Its beautiful production contained numerous illustrations, making it a very attractive 'picture diary of a new expedition in Sikkim'. Ernst Grob must have contributed most of the text. In the preface, Grob's brother declares:

> And over all suffering there lies always somewhere a peaceful reconciliation.

Sadly, in their case, the Tent Peak three were not able physically to meet again. Paidar and Schmaderer were interned with Heinrich Harrer and Peter Aufschnaiter at Dehra Dun (now Dehradun) in the foothills of the Himalaya. Paidar and Schmaderer also managed to get through the barbed-wire boundary fence. Heading for Tibet, they followed the Ganges into and beyond the Spiti Valley. At the village of Tobo they decided to purchase more provisions before crossing the border. Schmaderer went into Tobo but was stoned to death and his body left on the riverbed. Paidar subsequently gave himself up. Schmaderer's murderers were arrested. A few years later, in the eastern Alps, Herbert Paidar was hit by a falling rock and died.

1938-1939: the German Sikkim-Tibet expedition

Just as at heart Paul Bauer was a mountaineer, Ernst Schäfer was an explorer: Sven Hedin was his role-model and guide. As with Bauer, his first love became entangled with National Socialism to the extent that Hitler's *Reichsführer SS*, Himmler, agreed to sponsor Schäfer's third expedition, an exploratory investigation of Sikkim and Tibet. What was kept secret was the quest to locate evidence of the remnants of the 'Aryan' race and establish mutual economic and military arrangements with Tibet's rulers.

Schäfer had first intended to enter Tibet through China, but the war in the East blocked access from there. He next tried making overtures to the British in India to enter through Assam. The Government of India in Delhi and Hugh Richardson, the British representative in Lhasa, opposed allowing Germans through British India into Tibet. They used the genuine reason that Assam was unsafe, as Lightfoot's expedition against the Tibetans demonstrated.[2] Schäfer complained to Himmler of British intransigence. He then put pressure on various British friends of Germany. The British Ambassador in Berlin, Neville Henderson, a strong supporter of appeasement, lobbied for the Schäfer expedition. In the Oriental and India Officers' file on Schäfer it is recorded that Schäfer was backed by other influential people in Britain, including Lord Astor and Admiral Sir Barry Domvile, once head of British naval intelligence and a well-known supporter of Anglo–German friendship who had spent time in Germany with Himmler on shooting holidays. Domvile passed on Himmler's objections to Prime Minister Chamberlain, who desperately wished to avoid embarrassing Germany and soon gave permission for the expedition to proceed through Sikkim, where they were welcome to conduct their scientific studies.

Just before leaving London, Schäfer received a telephone call at his hotel from another of his heroes, Sir Francis Younghusband. It was not until he wrote a second edition of his book, *Fest der weissen Schleier* (*Festival of the*

---

2. In 1938, the British made a move to assert sovereignty over Tawang, on the Indo–Tibetan border in Arunachal Pradesh, by sending a small military column there under Captain G.S. Lightfoot. This expedition was met with strong resistance from the Tibetan government and local people; a protest was lodged against the British Indian government.

White Scarves, 1961) that he divulged their conversation about the problem of persuading the India Office to assist access to Tibet. Younghusband's advice was:

> Sneak over the border, that's what I should do, sneak over the border. Then find a way round the regulations.

Schäfer and his team completed the long journey to Gangtok by ship, rail and car. By 21 June, they were ready to move off up the Tista Valley into western Sikkim. First, they were invited by the *chogyal*, Sir Tashi Namgyal, to film the Phang Lhabsol festival and the war dance of the gods, dedicated to Sikkim's holy mountain, Kangchenjunga. The dancers, 'ecstatic, wild and terrible', waving swords, whirled around in front of the mesmerised audience below the distant sight of Kangchenjunga. (The visitors were fortunate to witness the festival, unique to Sikkim: it is performed once a year on the fifteenth day of the seventh month, to worship and give thanks to the mountain for unifying the nation during the thirteenth century, a reference to the Treaty of Brotherhood between Khye Bumsa, chief of the Bhotia (Tibetans) and Tetong Tek, religious leader of the Lepcha.) The deities of Kangchenjunga were invoked to witness this important event in Sikkim. (*Phang* means 'witness'.) The dancers wore red face masks with a crown of five skulls to represent the guardian deity. Later, Schäfer wrote of the festival that the gods who reside on Kangchenjunga were war-like, adding his view that:

> The Lepchas were not born to be leaders. Though they are good subjects, modest, hard-working and adaptable, they aren't fighters: they avoid any danger and that is why they are where they are, in the middle of the jungle, where they are in nobody's way.

The expedition film, *Geheimnis Tibet (Secret Tibet)* frequently alluded to the intellectually superior and physically robust ethnic groups whom they met. The film leaves the audience in no doubt about 'Aryan' inherent superiority. While Ernst Krause continued to film, Schäfer went hunting and Karl Wienert, the expedition's geographer, measured variations in the Earth's magnetic field. Bruno Beger, the anthropologist, spent his time luring locals into his field of study via the medicine chest. Once he had their confidence, out came his callipers, charts and measuring tapes to determine their 'type'. First, with his assistant, Edmund Geer, he wanted to try making a head-cast or mask. The only person available was Pasang Sherpa but he had a head injury. Irresponsibly, they pressed ahead, smearing plaster over Pasang's face. Normally, straws would be inverted up the nostrils but Beger merely wiped the wet plaster away from Pasang's nose. Pasang was having problems breathing; Beger could see he was frightened while he covered his eyes with plaster. He decided to continue and wait for the paste to set until he realised Pasang was having a fit. Gurgling and foaming through the plaster, he fell to the ground, his body writhing, sucking wet plaster into his mouth and nostrils while turning blue. Beger, a giant of a man, scraped the plaster off Pasang's face and rammed his huge finger down his throat to clear his airways: Pasang was saved. Apparently, as the wet plaster oozed into Pasang's nose the mountain god, Kangchendzonga, had appeared before him, seized him and shaken him violently. Pasang and the other porters located an image of a grotesque demon god and put it in front of Schäfer as a warning to unbelievers.

Schäfer was furious with Beger, knowing that if Hugh Richardson heard about this incident, they could all be ejected from Sikkim. Everyone concerned was bribed into silence, with the promise of no more mask-making in Gangtok.

The expedition proceeded north to Lachen, where they hoped to continue to the Tibetan border and follow Younghusband's advice to sneak over. They continued to within twenty-five kilometres of the border, where they occupied the Himalayan Club bungalow at Thangu. It was here that they had surprised Tilman, and later two more British mountaineers returning across the Himalayan divide from the 1938 Everest attempt. Peter Lloyd and Noel Odell had been looking forward to 'that well-remembered room with its great stone fireplace, round which we so often sat', but now there was a large European party in residence with tents in the garden over which fluttered an incongruous flag:

> The swastika on it was unmistakable, bearing a message so different from those scrawled on the Tibetan monuments.

The Schäfer expedition hunted *bharal* or 'blue sheep' for the museums of the Reich. They then continued to the Zemu glacier, finding untouched tinned food at Bauer's old base camp. They shot *argali* (wild sheep) and beautiful wild *kyang* (the Tibetan wild ass). Schäfer would cut the throats of some of the animals and drink their blood to give him strength and potency.

By chance, the key to entering Tibet by invitation, rather than sneaking over the border, presented itself with the visit in August 1938 of a member of the Sikkimese royal family who was living in Tibet. Schäfer presented him with gifts of western origin but also a large quantity of vegetables which the envoy said the Regent in Lhasa was fond of. The envoy was left in no doubt about the sincerity of Germany's friendship with Tibet and how much they would like to visit. The minister recommended Schäfer write a request letter: on 22 December, with a letter of invitation to visit Lhasa, they entered Tibet via the Natu La. By 19 January, the expedition had reached Lhasa. By May, after visiting Shigatse, the expedition reached Gyantse, to negotiate an urgent return to Calcutta as Schäfer desperately wanted to avoid internment should war break out. A huge amount of baggage was eventually flown out to Baghdad, then Athens, where they boarded a new German government aircraft, to be met at Berlin Tempelhof by an ecstatic Heinrich Himmler, who presented Schäfer with the SS skull ring and dagger of honour.

The unfeeling way in which Bruno Beger tried to achieve his aim in imposing a face mask on Pasang Sherpa, despite his obvious distress, indicates his lack of conscience. It is mentioned in *Himmler's Crusade* (2003) that Beger's work on face masks and taking measurements was filmed and is lodged in the Library of Congress. This work demonstrates that he inflicted pain on his subjects (although it has been edited out of the final film). Nevertheless, Schäfer's uncut footage showed:

> that violence underpinned the practice of physical anthropology; measuring bodies and making masks was not much different from possessing them or turning them into lifeless cadavers.

Beger was repeatedly interviewed about his SS activities after the war but managed to avoid execution. Hirt shot himself; Himmler took cyanide. Schäfer was interned by the Allied military government in 1945 but in June 1949 was exonerated of war crimes and released.

14

War, Partition and Reconnaissance

1940s: the new world order

This was a time of great turmoil. Significant changes in the balance of power across the world affected access to the Himalaya and Kangchenjunga.

Mao Zedong proclaimed the People's Republic of China on 1 October 1949. He quickly tightened his grip on China's southern borders with India, Nepal, Sikkim and Bhutan. To the south of the Himalaya, the British Empire was in full retreat, although there was no headlong rush for independence by the people of the Indian subcontinent because of intractable problems of race and religion combined with a wish to preserve some of the advantages of British colonial rule.

In partition year, 1947, Nepal found herself sandwiched between India, now an embryonic democracy, and Tibet, about to be overrun by a ruthless communist dictatorship. Changes in Nepal's international relations were being forced upon the ruling Rana regime.

Nehru, the Republic of India's first Prime Minister, had adopted a foreign policy of non-alignment, to be a bridge between the capitalist and communist worlds. Realignment replaced non-alignment, however, after the Chinese invasion of Tibet in 1950 and its incursions within Ladakh and into Assam had been discovered.

The Second World War hardly affected Sikkim, except for a decrease in tourists and climbers, leaving many of the Sherpas in Darjeeling out of work. Otherwise, the kingdom continued to function at its usual leisurely pace until the *chogyal*'s eldest son, Paljor, was killed in a plane crash. The accident left eighteen-year-old Prince Palden Thondup Namgyal as the future *chogyal* of Sikkim – and its last.

During the war, servicemen in India did manage to take short stretches of leave for trekking or to climb mountains. Short, low-budget journeys were made possible by the abundance of fifty-seven bungalows maintained by the Public Works and Forestry departments. These tourist rest-houses, mostly of stone, usually offered a fine view of the Kangchenjunga massif. Their red-painted roofs were noticeable from afar. They were sited within a day's walk of each other, so camping equipment was not needed. Any prospective journey had to be registered, which had the merit of avoiding over-booking. Until 1947, travellers into Sikkim had to apply one month in advance for dak bungalow passes from the Deputy Commissioner in Darjeeling. Frontier entry passes had to be obtained from the Political Officer in Gangtok. The former Himalayan Club secretary was no longer available in Darjeeling, so recruitment of porters was organised by Karma Paul, sirdar on all the Everest expeditions since 1924. Other famous Everest Sherpas such as Ang Tharkay began to take the initiative and offer their services to expeditions arriving in Darjeeling.

In September 1942, Group-Captain A.J.M. Smythe and Wing-Commander L.S. Ford climbed Lama Anden (5,868m), within a fortnight's round trip from Calcutta. This fast time was made possible by generally improved communications and recruitment efficiency of the Sherpa organisation.

A second ascent was made of Chomiomo (6,838m) by Harry Tilly with two brilliant Sherpas, Ang Tharkay and Dawa Thondup, in late July 1945. In October of the same year, Wilf Noyce reached the summit of Pauhunri (7,128m), with Namgar Sherpa and Ang Tharkay breaking trail. This was the second ascent after Kellas, but Noyce reached the summit only sixteen and a half days after leaving Delhi.

Despite political and social uncertainty following Indian independence, the Himalayan Club gathered strength again after the war had scattered its members. In 1949, Trevor Braham became Honorary Secretary and Charles Crawford (of Chomolhari fame) was elected President. Wilf Noyce restarted the invaluable *Himalayan Journal* after an interval of seven years. Colonel H.W. Tobin remained editor for the next nine years. The Club based itself in Calcutta; membership began to increase, with branches in Delhi, Bombay and Darjeeling. By 1950, Club membership had again reached 500, with many Indian members, particularly army officers who had recently taken up climbing as a sport. In 1951, Mrs Jill Henderson, whose husband ran a tea garden (estate) near Darjeeling, agreed to be the local secretary. According to Braham in *Himalayan Odyssey* (1974), relations with the Sherpas were never closer, with 'Tiger' badges being awarded for the best new Sherpa guides.

As Sikkim closed its borders, Nepal opened up. The first foreigners were allowed in as tourists in 1949. Kathmandu now became the gateway to Kangchenjunga; western approaches passed through Limbuwan, the country of the Limbu.

In 1952, after the invasion of Tibet, Braham described the changes taking place in Sikkim under Indian supervision. The police had become increasingly vigilant; the Indian army was 'everywhere'. The Indian government had appointed John Lall, an I.C.S. (Indian Civil Service) officer, as the new Indian *dewan* (prime minister) in Gangtok. He immediately set about modernising the state. In fact, on 14 December 1950, the *chogyal* signed a new treaty between India and Sikkim. (This period is comprehensively described by Andrew Duff in *Sikkim: Requiem for a Himalayan Kingdom*, 2015.) The treaty allowed Sikkim to 'enjoy autonomy in regard to its internal affairs'. The government of India, however, would remain 'responsible for the defence and the territorial integrity of Sikkim' and enjoy 'the right to station troops there'. The government of Sikkim, moreover, was to have 'no dealings with any foreign powers'.

Thondup became Regent after his father had withdrawn from society to concentrate on religious contemplation. His problem in preserving the country's heritage was to give a voice to the indigenous inhabitants, the Lepcha, the Limbu and the Bhotia (see Chapter 2). All three ethnic groups co-existed reasonably well. Problems had started with the influx of Nepalis escaping Rana tyranny during the latter half of the nineteenth century to take advantage of the Indian government's land-grants. Before long, these Hindu farmers and tradesmen were to outnumber the combined Sikkimese ethnic groups. In a democracy, the Hindu Nepali would predominate: eventually, they voted in favour of Sikkim becoming the twenty-second federal state of India in 1975.

By the early 1950s, Sikkim was changing fast, as Trevor Braham noted in 1952 on his final visit. The *dewan*, John Lall, for example, was refusing to permit travel north of Thangu: the Indian government was already apprehensive about the Chinese, despite Prime Minister Nehru's overtures to Chairman Mao. Even greater restrictions were imposed on visitors to Sikkim's mountains. Braham reported that the Himalayan Club hut in the Jha Chu Valley had been vandalised by local herdsmen and was barely habitable. The Indian army was soon to cause immense damage to the environment by driving roads to the northern borders, cutting down swathes of forest for cooking-fires and for heating road-surfacing tar. Some of the trees cut down were over 500 years old.

Nepal, from 1949 to 1950, after years of isolation, changed course by allowing aid workers, mountaineers and a few tourists into the country. The centre of Himalayan climbing from then on began to shift from Darjeeling to Kathmandu. In just about every sphere of life, Nepal had stagnated since the establishment of the Rana regime in 1846. Nepal had to start from scratch since it lacked even those basic amenities that most countries take for granted, such as roads, a health service, child and adult education, justice for all, fair taxation. In a bloodless revolution, with a constitutional monarchy established under King Tribhuvan Shah, everything was set to change. The king died in 1955 without fulfilling his promise to set up a constitutional assembly. His son, King Mahendra, held Nepal's first election, giving hope to the people; those hopes were not fulfilled: in 1960, the king, with the support of the army, dismissed the government and initiated thirty years of the cumbersome and autocratic *Panchayat* system.[1]

Nepal's only real success was tourism: the ascent of Everest in 1953 put the country on the map as a global tourist destination. The first wave of visitors walked into a time warp of old ways and values. The ready, unaffected smile and the hospitality offered to complete strangers disarmed and charmed the new visitors. Word got out about Nepal's likeable, hospitable people and astounding mountain landscape.

1949: the Dittert Swiss expedition

At approximately the same time in 1949 as Tilman was travelling through the central Nepal Himalaya, the Swiss, led by René Dittert, had established themselves in the north-east of Nepal on the west and north sides of the Kangchenjunga massif. The other team members were Madame Annelies Lohner and Alfred Sutter, who had been with Dittert on a successful climb in the Garhwal in 1947; the final member was Dr Wyss-Dunant of Geneva. They were accompanied by two Grindelwald guides, J. Pargatzi and A. Rubi.

They set off from Darjeeling on 1 May after Pasang Lama had enrolled 200 porters to shift 4,000 kilos of equipment, food and fuel to Base Camp at Lhonak by 19 May.

After a few days' rest, Dittert split the team into reconnaissance groups: Mme Lohner, Stutter and Rubi went north towards the Chabuk La to investigate the mountains within the triangle of country defined by Drohmo ($c.6,881$m), Jonsong Peak ($7,462$m) and Nupchu ($c.7,030$m). Dittert, Pargatzi and six Sherpas headed south then south-east to the Ramthang glacier,

---

[1] *Panchayat* was a political system without parties. Under direct rule, King Mahendra introduced a four-tiered structure: village, town, district and national *panchayat* (council).

to explore a way up Kangbachen (7,903m). After three days, having camped at around 5,500m, they could find no safe way and retreated to Lhonak.

Dittert now concentrated on the west wall of the main north ridge of the Nepal–Sikkim frontier. The team focused on climbing the Sphinx and Pyramid peaks, pitching Camp 3 around 21 November on the Langpo La. By then, the guide Rubi, 'much to our regret had to go back to Europe suffering from insomnia, a badly sunburnt face and nervous exhaustion,' wrote Dittert in his account for the *Himalayan Journal* (1950–51).

After a period of stormy weather, the team put up Camp 4 on the north-east spur of the Sphinx at 6,700m. Pargatzi and Dittert, on 6 June, climbed to its summit (6,979m), first reached in 1926 from Sikkim by Cooke, Chapman and Harrison, accompanied by their Sherpas, Gyalgen, Ajeeba and Dawa Thondup.

Dittert, Paragtzi and the Sherpas had to descend just over 100m in order to reach the col from where they could start the final climb to Pyramid Peak (7,164m) via its north-east arête. After many difficult cornices which forced them off the arête, they reached their high point:

> We stood on the summit crest about eighty metres horizontally from the top. We could hardly see it and were separated from it by a barrier of vertical spikes of frozen snow and ice. We would have had to have broken them down, one after another ... It would have been a senseless undertaking as anyway the Pyramid had been conquered.

They were overjoyed with their effort, despite the mist and stormy weather at over 7,000m – but with still a kilometre of arête to re-negotiate before climbing into their tents.

After four days' rest, Pargatzi and Dittert tried unsuccessfully to find a way up Drohmo (*c.*6,881m), a peak lying between the Kangchenjunga and Broken glaciers, first investigated by Kellas and named by him 'Longridge'. Sutter had explored the north and north-west sides and thought them unclimbable. Dittert and his guide could not find a way up Drohmo. They gave up and climbed Tang Kongma (6,250m) instead. They established Camp 2 at 5,488m and started at 6 a.m. On the first rope were Stutter, Mme Lohner and Pargatzi; on the second were Dr Wyss-Dunant, Pasang and Dittert. They ascended snowfields, then crossed a sérac and crevasse to reach the summit by 10 a.m. Unique views of the north-west flanks of Kangchenjunga's main summit, sixteen kilometres to the south-east, were revealed.

The party then left the Kangchenjunga glacier for the Nepal–Tibet frontier range, hoping it would be less affected by the monsoon. They accomplished some useful exploration of the Lhonak and Tsisima glaciers and, after crossing into Tibet, explored the Chabuk glacier on the north side of the Himalayan divide. They attempted Nupchu (c.7,030m) but gave up 250m from the summit. They climbed Dzanye (6,341m) on 30 June. All the team summitted, including all the Sherpas, bringing a most successful mountaineering expedition to an end.

For the walk-out, the Swiss split up, with Dr Wyss-Dunant taking the direct route to Darjeeling, while Dittert followed a longer route leading over the Nango La (4,726m) from Ghunsa. They followed the Zari Chu Valley for several days, then descended the Tamur Valley to bring them to Taplejung, the chief town of Dhankuta province. From there, they trekked through park-like countryside, walking mainly on a carpet of pine needles, past Angbung, and after five days arrived at the frontier on a good road leading to Sandakphu. By 7 August they had returned to Darjeeling to meet Dr Wyss-Dunant on the ninety-ninth day of their Himalayan adventure. This group constituted not just the first mountaineers to be officially invited into Nepal, but the first to have freedom to roam there since Sir Joseph Hooker in 1849, a hundred years before.

John Hunt was asked, following his return from Everest in 1953, what was there to climb, now the highest place had been reached? 'Kangchenjunga,' he replied, because:

> Those who first climb Kangchenjunga will achieve the greatest feat in mountaineering. It is a mountain which combines not only the severe handicaps of wind, weather and very high altitude but technical climbing problems and objective dangers of an order even higher than those we encountered on Everest.

Had the Swiss climbed Everest in 1952, the British would have gone for Kangchenjunga in 1953 as the next best thing. Since K2 had been climbed in 1954, Kangchenjunga became the main Himalayan objective, being the highest unclimbed peak in the world.

By 1950, the problem of access to Kangchenjunga had changed. Before the war, the customary approach had been through Sikkim. Less serious investigations had started with Freshfield's encircling of the massif in 1899, Kellas probing up the Zemu glacier and the north-west of Sikkim, and Grob and his friends Schmaderer and Paidar looking for an alternative route in

1936 and 1939. Reginald Cooke, with John and Joy Hunt, explored the Zemu glacier region in 1937: Cooke and his Sherpas reached close to the North Col of the north ridge. Apart from the Bauer route, all the other possible routes on the eastern, Sikkimese side had been dismissed as too dangerous or logistically impossible.

The forbidden west side of the massif had been encroached upon (see Chapters 7 to 13, *passim*), though most visitors had failed to obtain Nepali permission. Frank Smythe, one of the experienced alpinists on the 1930 attempt, was particularly dubious about anyone climbing Kangchenjunga from the Yalung. Smythe was on his first Himalayan expedition and had not explored anywhere near the south-west face, other than to examine it by telescope from Darjeeling, yet he writes of Kangchenjunga:

> [the] South-west face is, in effect, not possible as the distances and exposure to the west wind count against even getting to the snowy shelf seen from Darjeeling leading up in a westerly direction.
>
> The snowy shelf looks, and probably is, desperately dangerous owing to falling stones and avalanches, and its dangers must be considerably increased by its southern and consequently warm aspect.

Smythe concluded:

> There would seem little justification for a further attempt from this side.

Smythe's opinion mattered; certainly, his comments did not encourage mountaineers to attempt the south-west face of Kangchenjunga in a hurry. With the eastern aspect closed by the Indian army, the only option remained the western side of the mountain. The Nepalis saw themselves in a similar position to Switzerland, mountainous and multilingual, sandwiched between powerful neighbours. This was their chance to become a similarly neutral, rich, tourist destination.

1951: Frey and Lewis
In the autumn of 1951, George Frey, the assistant Swiss trade commissioner in Bombay (now Mumbai), managed to obtain permission to visit the Yalung side of Kangchenjunga with his British friend, Gilmour Lewis, a Welsh mining expert. He was a noted amateur climber in Switzerland.

Frey had written to his fellow Swiss Günter Dyhrenfurth for mountaineering suggestions. The recommendation was to take:

> a close look at the upper Yalung glacier, having reached the conclusion years ago, that Kangchenjunga's south-west front, with its perhaps undeservedly ominous reputation since the 1905 catastrophe, required further study.

The next letter Dyhrenfurth received from Frey was written from Darjeeling on 22 September, a few days before Frey's departure with Lewis and a fine team of Sherpas, led by Tenzing Norgay. That was the last Dyhrenfurth heard from Frey until November 1951.

Frey, Lewis and their Sherpas crossed the Semo La and climbed a 5,488m rock peak from Tseram in the area marked Kangla Nangma on the Marcel Kurz map. They then moved up the Yalung Valley and glacier, using the same campsites as Raeburn and Crawford. Frey and Tenzing visited Pache's grave while Lewis and Ang Dawa crossed to the western rib on Talung Peak.

Lewis, who was suffering from jaundice, left Yalung (to return to work) by making the second crossing of the Rathong La. Lewis carried on to Darjeeling while Frey remained at Dzongri (or Jongri) for further climbs. A few days later, on the morning of 29 October, Frey, with Tenzing and Ang Dawa, attempted a 5,793m peak about a kilometre south-east of Kokthang. They were following a steep gully of snow and iced-up rocks. Tenzing stopped to put on his crampons; Frey continued leading. He then fell, past Tenzing, who tried to stop him (breaking his finger in the process), for 430m. The Sherpas carried Frey's body to the end of the glacier and laid it by a huge block of stone. They built a cairn and placed his ice axe over it.

Frey and Lewis had concluded that the Yalung face of Kangchenjunga was not impossible and would certainly benefit from further exploration by a larger, better-equipped party. So convinced was Lewis that he was to return twice to reconnoitre the face, in 1953 and 1954.

1953: Lewis and Kempe

In April 1953, Lewis returned to the Yalung Valley with John Kempe. Lewis had been captivated by the area and was intrigued to know if a safe route could be found above the upper Yalung glacier.

Lewis approached John Kempe, who since 1951 had been the founding principal of the Hyderabad Public School in central India. John Kempe (1917–2010) was born in Nairobi. He was educated at Stowe and then Cambridge, where he read Economics and Mathematics. In the Second World War he flew Spitfires and Mosquitoes and was promoted to Squadron Leader, being twice mentioned in despatches. He was lucky to survive nearly seven years of aerial warfare.[2] After the war, Kempe enjoyed three seasons' climbing in the Alps and then, in 1952, joined Harry Tilly and John Jackson for a climbing expedition to Nilkanta (6,597m), on which they were unsuccessful, owing to heavy falls of snow.

Lewis and Kempe left Darjeeling in the spring of 1953, to spend April and May around the Yalung glacier. They first attempted Buktoh (5,976m) on the west side of the valley. They also attempted Kokthang (6,147m) and Kabru North Peak, but gave up owing to worrying snow conditions just below each summit. From high on Kabru, and from what they could see of the south-west face from lower on the surrounding valley sides and the glacier, they thought there might be a reasonably safe route up the south-west side of the mountain. Their positive report and photographs resulted in a second reconnaissance, this time with the blessing of the Royal Geographical Society and the Alpine Club who, between them, funded those members travelling to India from Britain; this expedition was otherwise privately organised and funded.

1954: Kempe's reconnaissance

The team which sailed for India in March 1954 consisted of John Kempe (thirty-seven); Gilmour Lewis (thirty-one); Trevor Braham (thirty-two), who had experienced many Himalayan climbing seasons, mainly in Sikkim); and Ronald Jackson (thirty-seven). Jackson came from Burnley; he was on his first Himalayan visit, although he had climbed in the Alps and reached a good standard on British rock. Other members of the party included John Tucker (thirty-one), a teacher with several Alpine seasons who had been placed on the Everest 1953 reserve list, and the expedition doctor, Donald Matthews (thirty-seven), a New Zealander with no climbing experience.

The Joint Committee of the Royal Geographical Society and the Alpine

---

2   Kempe went on in 1968 to become headmaster of Gordonstoun, where this editor was one of his pupils. He never participated in the expeditions run by the school's Mountain Rescue Unit, nor did he ever (to my recollection) talk of his Kangchenjunga experiences. He did introduce some (at the time) unpopular liberal reforms, such as the introduction of female pupils, thereby consolidating his link with the school's founding philosophy developed by Kurt Hahn and Geoffrey Winthrop Young.

Club had applied to HM Government of Nepal for permission to climb on Kangchenjunga in 1955 and 1956. The Committee received a positive reply during the Kempe reconnaissance. In view of the impending 1955 expedition, Kempe considered which would be the best of the three approaches to the Yalung glacier in March, given likely logistical problems arising from how the weather affected the trail and road surfaces.

The high road via Pemayangtse to Dzongri and over the Rathong La (5,122m) would be in deep snow, but food would be available for porters as far as Yuksom. The second possibility was to trek from Pemayangtse northwest up the Rimbi Chu, by way of Danphebir (4,633m) and over the Gareket La. Again, late winter snow could cause huge problems for the porters. The third alternative became Kempe's recommendation, which was to travel from Darjeeling for three days as far as Phalut, with dak bungalows and food available along this classic route. Beyond Phalut lay a low-level route into Nepal via Khebang and Yamphodin, to cross a saddle at 3,050m near the Semo La and so down to Tseram. There should be no winter snow problems and food would be available, assuming a sirdar and one expedition member could make advance arrangements with the village headman. Lewis, on his return journey to Darjeeling, followed this route, checking food stocks as he proceeded.

This reconnaissance did look at climbing Kangchenjunga via the southwest face, but where exactly the route should go remained inconclusive. Three possibilities were looked at: to the left of the Great Icefall and the Western Buttress; to the right through the glacier basin below the Talung Saddle and to the east of the lower icefall, or up a rock feature which became known as Kempe's Buttress. Kempe and his team preferred this route, although there were constant worries about snow conditions, icefall, sérac collapse and huge avalanches.

Trevor Braham later reflected that their 1954 reconnaissance should have attempted to reach a higher viewpoint on the slopes opposite the southwest face to discern a practical way to the Great Shelf. Kempe's report was presented to the sub-committee of the Alpine Club. Although inconclusive, Kempe's final advice, if conditions above the Rock Buttress were unstable, was 'Pache's grave route is certainly worth further investigation'.

The most important result of the 1954 reconnaissance was the realisation that Aleister Crowley's instincts about the best way up the south-west face were correct. In John Tucker's *Kangchenjunga* (1955), his account of the 1954 expedition, he writes:

Is it possible in the light of our reconnaissance to dismiss Crowley's excessive optimism as springing from a lack of technical knowledge? Guillarmod's defeatist attitude may well have contributed to the long-standing neglect of Kanchenjunga's West Ridge and South-West Face. It must also be conceded that Crowley's route up the steep slopes towards the Kangbachen Peak (one of the lower summits of the Kanchenjunga West Ridge) was not ill-chosen. Kempe, the leader of the 1954 reconnaissance, has made the following observations, 'The bottom part of this route is to some extent threatened by ice-cliffs from above, but the danger of passing below them is not unduly great and only one large avalanche was observed by us. After the first thousand feet there do not seem any objective dangers.'

We see in hindsight that the Great Beast had all the instinct of an incisive great pioneer climber and that Guillarmod's judgement was flawed. The general line of a route to Kangchenjunga's summit had been discovered.

PART 4
Ascents

15

The First and Second Ascents

1955: the first ascent

The Alpine Club sub-committee of Sir John Hunt, Douglas Side and Mike Westmacott recommended to the Joint Committee (of the Alpine Club and the Royal Geographical Society) that a party, led by Charles Evans, 'Shall make a strong reconnaissance of Kangchenjunga in 1955.'[1]

In the spring of 1954, Charles Evans was cabled by Sir John Hunt as he was returning from Kathmandu:

> To Evans or Hillary: HMG [the government of Nepal] has given the [Joint] Committee permission for climbing Kangchenjunga in 1955 and 1956. Will one of you lead a reconnaissance next year? Kempe report is inconclusive and climbing route is far from solved.

Evans confirmed he would lead the 1955 expedition. He enquired if the financial support was sufficient for oxygen to be brought: he had, from the start, envisaged an in-depth reconnaissance.

The main challenge was the sheer scale of the physical problems to be overcome. The highest anyone had reached was believed to be about 6,000m, by three of Crowley's team, Reymond, Pache and Salameh, on 1 September 1905. The 1955 expedition would have to negotiate some 3,000 vertical metres of largely untrodden mountainside. It was a huge, fascinating challenge for which Charles Evans was absolutely the right leader.

Evans was sent to an English preparatory school and then, on a scholarship,

---

[1] Douglas Side (1896–1961) was the first Honorary Secretary of the Mount Everest Foundation and a member of the Alpine Club Committee from 1954 to 1956. Mike Westmacott (1925–2012) was a member of the 1953 Everest expedition and Alpine Club President from 1993 to 1996.

to Shrewsbury School. At University College, Oxford, he switched from Classics to Medicine. By then, he had climbed in all the main upland areas of Britain and managed a season in the Alps. He qualified as a doctor in 1943 and was posted to India then Burma as medical officer to an infantry battalion. There, he taught himself Hindi and managed the occasional climb, such as Kinabalu. After the war, he was appointed surgical registrar in Liverpool. Jim Perrin wrote in his excellent obituary of Evans:

> He must have been well regarded since he spent more time in Nepal than in a Merseyside operating theatre.

Norman Hardie, Evans's deputy, had climbed with Evans. He wrote about him in *On My Own Two Feet* (2006):

> We all felt humbled by this man's quiet authority, his wit and his tremendous prestige amongst the Sherpas.

The Joint Committee gave Evans a free hand to choose his own team: Evans asked Dawa Tenzing, whom he knew well from three expeditions together, to choose the best Sherpas from Solu Khumbu. This team was modest in number and demeanour. They were all new to Kangchenjunga.

Norman Hardie (aged thirty) came from New Zealand; he was a very experienced, distinguished ice-climber. Hardie took on responsibility for oxygen and surveying.

Tony Streather (twenty-nine), an army captain, had accompanied the Norwegian Tirich Mir expedition as transport officer. He reached the 7,692m summit wearing a golfing jacket, without an ice axe and with a bedding roll weighing over five kilos! He reappeared as transport officer to the American 1953 K2 expedition and climbed above 7,600m without too much trouble. His ability to speak Hindustani was of great value; he was put in charge of the porters.

George Band (twenty-six) had been the youngest member of the 1953 Everest expedition; illness had prevented him going high. After Everest, he reached 6,340m on Rakaposhi. He undertook the thankless job of trying to satisfy everyone's food requirements.

Tom McKinnon (forty-two), a pharmacist from Glasgow, was the oldest member of the team. He was known for his great stamina, which had previously been demonstrated in the Himalaya. McKinnon was the expedition's stills photographer.

Neil Mather (twenty-eight), a Manchester textile technologist with a good record of Alpine snow and ice climbs, was also a prodigious long-distance walker.

The expedition doctor was John Clegg (twenty-nine), a Liverpool University anatomist. His diverse experience included being a Territorial Army paratrooper and a competent all-round Alpine climber.

John Jackson (thirty-four) was a forthright Yorkshire teacher. He spent the war at the Aircrew Mountain Centre at Ganderbal and Sonamarg; this allowed him to climb in Kashmir, the Garhwal and Khumbu, where he participated in a well-publicised hunt for the Yeti.

The youngest member was Joe Brown (twenty-four). He had no Himalayan experience but was Britain's most formidable rock-climber and alpinist. He was a Mancunian plumber and builder with a great sense of humour. Joe was a popular expedition member; it seemed that Evans had chosen well, with every member commenting on the team's cohesive nature. Joe was lampooned (because of his apparent social incongruity) by his mates at work and in the Rock and Ice climbing club, but perhaps most memorably by the redoubtable Tom Patey in 'Red Pique' (The Alpine Club Song, from *One Man's Mountains*, 1971):

> Our climbing leaders are no fools,
> They went to the very best Public Schools,
> You'll never go wrong with Everest Men,
> So we select them again and again,
> Again and again and again and again.
> You won't go wrong with Everest Men,
> They went to the very best Public Schools,
> They play the game, they know the rules.
> Customs change and so alas
> We now include the working class,
> So we invited Good Old Joe
> To come along and join the show.
> He played his part, he fitted in,
> He justified our faith in him.
> We want the climbing world to know –
> That the chaps all got on well with Joe ...

The sirdar, Dawa Tenzing, had twice carried loads to Everest's South Col without using oxygen. He was forty-five, still very strong, with a natural dignity. He chose Annullu, the first Sherpa (in 1953) to reach the South Col, as his deputy. Evans requested thirty Sherpas in total; fifteen were experienced enough to climb high. In addition, the expedition required more than 300 porters to carry about 6,000 kilos of equipment, fuel and rations from Darjeeling to Base Camp, a walk-in of more than 100 kilometres.

It was generally recognised by 1955 that pre-monsoon was the most favourable season for climbing high in the Eastern Himalaya, so the team assembled in Darjeeling, ready to leave by 14 March.

Charles Evans had been summoned to Gangtok to discuss the operating conditions for the expedition to climb Sikkim's sacred mountain. Even climbing from the Nepali side is considered sacrilegious and liable to disturb the resident protective deities. Shortly before leaving England, Evans had been informed that the Sikkim government objected to any attempt to climb the mountain. To try to find a solution, he drove to Gangtok, where he was met by the Indian Political Officer, Mr Apa B. Pant, who then introduced him to the *dewan*, Mr Nari K. Rustomji, to assist in introducing Evans to the *chogyal* of Sikkim, His Highness Sir Tashi Namgyal and his heir, Thondup Namgyal.

Evans writes in *Kangchenjunga: the Untrodden Peak* (1956):

> I promised that we would leave the top and its immediate neighbourhood untouched and would go no further up the mountain than was necessary to assure us that the top could be reached.

According to Andrew Duff in *Sikkim: Requiem for a Himalayan Kingdom* (2015), sixty years later, Evans acknowledged that the summiteers stopped a few metres short of the top and received praise for showing respect. But Nari Rustomji (who was involved in the negotiations) tells a different story, that the Palace granted permission for the expedition to 'proceed only as far as necessary to ascertain whether or not there was an approach'.

When the party claimed they had approached the summit to within touching distance, Thondup was 'enraged, convinced that this was a deliberate breach, in spirit at least of the solemn undertaking given by Evans'.

Thondup's moods could quickly take a dark turn, not helped by overconsumption of various prescription drugs. His outburst concerning the sanctity of Kangchenjunga's summit should be seen in the context of his loyalty to his country – at the time under threat of absorption by India – and his status within it.

Evans had been unable to obtain permission to journey through Sikkim. The party therefore left Darjeeling bound for Mana Bhanjyang. Not even casual visitors were hiring help for their journeys into the Kangchenjunga massif or anywhere within the 'Inner Line'.[2] Fearing the Chinese, the Indians had shut off access to huge areas of the most wonderful landscape on the planet.

The expedition, in any case, took Kempe's reconnaissance advice and travelled along the more southerly route to enter Nepal since they were approaching early in the spring season. From the roadhead, they followed the Singalila Ridge through Sandakphu to Phalut, as many tourists had done, but from there, with newly acquired permission, they plunged into deep, heavily vegetated valleys on the western side of the massif. They descended some 2,000m from where, during the rest of the walk towards Tseram, they passed out of forest to hillsides of intense terrace cultivation with little shade. Most of the walking was completed in the early morning, enlivened by unusual birds and colourful butterflies.

The porters were paid off at Yalung Baro, beyond Tseram and at 4,000m well up the Yalung Valley, an ideal place to begin acclimatisation.

Band, Mather and Jackson crossed the Mirgin La to Ghunsa, not only to acclimatise but to order Sherpa food of *atta* (wheat flour) and *tsampa* (barley flour). Meanwhile, loads were being carried up the broken Yalung glacier through Moraine Camp, Crack Camp and Corner Camp to Base Camp. It was not easy going, as the loose moraine was frequently covered in fresh snow. Hardie, Brown and Jackson left to climb Kokthang (6,147m) but failed in poor weather; they did, however, ascend two snow peaks to the south-east.

Base Camp was eventually established at the foot of the Lower Icefall. George Band was the first to occupy it, on 12 April, with two Sherpas. The next day, it snowed continuously. He remarked:

> I was able to spend the first twenty-four hours ticking off each avalanche on the tent frame with a pencil stub. At the end of the day there were forty-eight ticks, and since I had been asleep for a third of the time this gave a frequency of one avalanche every twenty minutes.

Following suggestions made by the 1954 reconnaissance, Band and Hardie climbed Kempe's Buttress (although the climb had first been led by John

---

2   The 'Inner Line': a protected area requiring an entrance/travel permit by the Indian government to regulate movement near Indian international borders.

Jackson). After fixing 150 metres of rope, they pitched a tent at the top of the buttress. Their intention was to find a way through the nearby icefall to reach the Upper Icefall leading to the Great Shelf. This (lower) icefall was, according to Band and then Evans, much more difficult and dangerous than the Khumbu Icefall on Everest: it was unsuitable for load-carrying Sherpas. This route was abandoned in favour of a gully Hardie had spotted, to the left of and above Pache's Grave. This new route necessitated the whole expedition moving Base Camp to just below the site of Pache's Grave at 5,500m.

Band and Hardie put in the new Camp 1 at 6,000m. They explored beyond it to find a way over the 'Hump' at the top of the western rock buttress as far as the Upper Icefall, where a safe site for Camp 2 was found: the terrain was rounded enough for any avalanches or snow sloughs to bypass it. Hardie and Band then descended for a well-deserved rest.

The next stage required an Advance Base Camp, Camp 3 (6,650m), halfway up the Upper Icefall. Brown and Evans established it; it was then stocked with 1,500 kilos of stores. Evans and Hardie, accompanied by Annullu, carried on to establish Camp 4 at 7,165m. The following day, 13 May, was memorable: Evans and Hardie, in poor weather, managed to cross the Great Shelf to a site for Camp 5 at 7,710m, just below the Gangway, the equivalent of reaching the South Col on Everest.

At Base Camp, resting before the push for the summit, the leader, without fanfare, came into the mess tent with a mug of tea and outlined his plans. Everyone was allocated a vital role. The first summit pair would be Band and Brown; the second, Streather and Hardie. Norman Hardie summarised Evans's plans in *On My Own Two Feet*:

> I think the upper part of the mountain will be difficult ... The first assault pair will have a night there (Camp 6) and next morning go as far as they can. If they reach the top that will be fine. They are to ... descend to pass Camp 6, converse there with the next pair, and then they will escort the second pair's two Sherpas, taking them down to Camp 5 that afternoon. On that same day the second assault will go up to Camp 6, settle in and get the report of the first party as they pass on down. Next morning they will make their attempt. George and Joe will be the first pair, with Norman and Tony next.
>
> There was a surprising silence for about five seconds, then laughter and congratulations. I was happy to be officially included but uncomfortable about Charles not being in one of the pairs.

The build-up was interrupted by bad weather, extreme tiredness and one severe case of snow-blindness, but by 22 May, Camp 5 was occupied and a rest day taken before the push came on 24 May to establish Camp 6 at about 8,230m. Fortunately, the snow on the Gangway was firm. Evans, Mather and Dawa Tenzing led in turn on the first rope so that Band and Brown could conserve their energy. Everyone was on oxygen and carrying loads of around twenty kilos. At 2 p.m., the party reached an outcrop of broken rocks at 8,200m and hacked out a ledge in the forty-five-degree slope. They soon hit rock; one third of the summit tent draped over the edge. Band and Brown drew straws as to who should sleep in the outside position: Band lost. They drank lemonade, then a mug of tea followed by supper of asparagus soup from a packet, a tin of lamb's tongue with mashed potatoes and a nightcap of drinking chocolate. They slept in their bags with every stitch of clothing, including their boots, to avoid them freezing solid as Hillary's had done at his Everest top camp.

May 25 dawned fine. At 8.15 a.m., after two pints of tea, with biscuits, Band and Brown left for the summit. They knew they had to move off the Gangway to the right, to a series of snow patches up the rocky summit cone. They turned off too early and had to reverse, wasting a precious hour. A favourable line was then found: at first, they belayed each other; Brown was obviously going better and so stayed in the lead while putting in the occasional piton to safeguard against a fall. They arrived at the pinnacles on the west ridge and stopped to take in lemonade, toffees and mint-cake. It was 2 p.m., leaving only two hours of oxygen to cover their return past Camp 6 to allow room there for the second pair. The going became easier until Brown decided to climb a crack about six metres high. From the top, he yelled down that the summit was only six metres away and two metres above them. They stopped at the foot of a gentle snow cone which was the summit. Out of respect for the Sikkimese, they went no higher. It was 2.45 p.m. After the summit photograph and orientating themselves, they descended. Extreme tiredness set in after discarding their now empty oxygen cylinders. Band fell when a patch of snow gave way but arrested himself with his ice pick.

As darkness fell, they reached their tent, to be greeted by an anxious Hardie and Streather. Band and Brown should have descended to Camp 5, but darkness and tiredness necessitated the four men crowding into the two-man tent hanging over its ledge. Somehow, they passed the night, with Band back in the 'dog-house' since he was used to it!

The next day, Band and Brown descended as Hardie and Streather repeated the route to the six-metre crack, which they managed to avoid by skirting around it on snowed-up rocks, then stopped just below the summit at 12.15 p.m. Streather, owing to a dropped cylinder, made the descent without oxygen. They too spent a second night at Camp 6.

By 28 May everyone had come off the mountain, well pleased that the Kangchenjunga 'reconnaissance' had succeeded in climbing it! The team's jubilation was curtailed: Pemba Dorje had returned from a high carry, exhausted. After apparently recovering, and despite all John Clegg's ministrations, he died of a cerebral thrombosis on 26 May just as Hardie and Streather reached the summit. Dorje was buried near Pache's Grave, under a rock inscribed '*Om Mani Padme Hum*'.

Pemba's death was the one unhappy event on the whole expedition; otherwise, it had been an enormous success. Sherpas and climbers, firm friends, had shared the success equally.

1977: the Indian army ascent of the north-east ridge

The second ascent of Kangchenjunga began badly when the leader, Colonel Narinder Kumar, was received by the *chogyal* of Sikkim. On broaching the subject of a joint Indo–Sikkimese expedition to Kangchenjunga, the *chogyal* stood up and declared, according to Kumar in his *Himalayan Journal* article of 1978, 'How dare you climb the mountain we worship?' Kumar proposed that, like the British, they would leave the top two metres untrodden. With biting sarcasm, the royal response was:

> I have heard that one before. The tallest Britisher who got to the summit was six foot three inches and his head was above my God's head.

The *chogyal* then invited Kumar to witness the Pang Lhabsol festival at Tsuklakhang monastery in honour of the deity of Kangchenjunga.

Kumar gave up his idea until Sikkim became part of India in 1975. Shortly after his successful Trisuli Ski expedition, he met the Chief of the army staff, General Raina, who inquired why the army had never launched its own mountaineering expedition, despite deployment at high altitudes. Kumar suggested that climbing Kangchenjunga from the east was one of the biggest challenges left in climbing. General Raina supported the idea, leaving Kumar to select the most experienced climbers in the Indian army.

Narinder Kumar was born in Rawalpindi in 1933. He and his three brothers all joined the Indian army after Partition, the family having moved to Shimla. He became the first Indian to reach 8,750m on Everest, as well as having climbed Nanda Devi, Chomolhari and Kamet.

In 1976, Kumar chose Major (later Colonel) Prem Chand as his deputy and climbing leader. Chand was born in 1942 in Lahaul and Spiti. He was a specialist in winter warfare. Later he worked in mountaineering institutes across the country, influencing among others Bachendri Pal, the first Indian woman to climb Everest, through training; he was also active in the conservation of Himalayan flora and fauna and indigenous culture, especially in relation to remote Buddhist communities.

Naik Nima Dorje Sherpa had also climbed above 7,600m. Known as 'N.D.', he had climbed Kamet, Bander Punch and Trisul and made the first ascent of the highest peak in the Kishtwar Himal, Sickle Moon. Over twenty members were selected after a training-cum-selection climb in the Zemu glacier area in the late autumn of 1976. Fifteen Sherpas from Nepal and six from Darjeeling, six high-altitude porters from Manali and a further support party of ten Ladakhi scouts completed the team.

There were significant differences on this Indian expedition to the northeast spur from the first two attempts by the Austro–Bavarians in 1929 and 1931. The most obvious was the huge increase in party size, with over four times as many porters and climbers in 1977 than in 1931. This increase derived in part from the decision to use oxygen which, for ethical reasons, Paul Bauer had rejected.

The Indian army team decided to approach the mountain during the pre-monsoon period, when it was assumed powder-snow conditions would be more favourable. The opposite applied to the walk-in from Lachen to the Green Lake base camp. Deep snow covering a tangle of jungle vegetation produced desperate conditions for porters. Strikes were so frequent that the army had to call in the Indian air force to helicopter food supplies up to 5,000m. The helicopter was also used later to 'medevac' members suffering from oedema.

The line up the north-east spur to its crest followed the Bauer expedition route, with similar problems. At about the same place where Schaller and Pasang Sherpa fell, the Sikh climber Sukhwinder Singh (who had married just before joining the expedition) fell while traversing a fixed rope, became entangled and broke his neck.

Seven other members were confined to their tents with injuries or high-altitude sickness. On the ridge, those still fit hacked away at tunnels through the ice towers, as had the Austro–Germans. The weather remained bad. Everyone worked at pushing out the route, with Prem Chand and N.D. Sherpa leading the way. At the supply depot, loads for the high camps were accumulating. The supply team, however, reported that some fifty porter loads had disappeared in a monster avalanche and could not be salvaged.

Thanks to Kumar's strong leadership, the expedition continued to make progress despite various setbacks: on 24 May, Camp 6 was established at 7,630m. Another problem arose when the Sherpas sent to ferry to Camp 7 for a summit attempt abandoned the mountain, unwilling to proceed along a wafer-thin snow arête 800 metres long. N.D. Sherpa and Tashi Dorje, however, made the arête safe with a fixed rope and Camp 7 (8,003m) was established on 30 May by N.D. Sherpa and Prem Chand. The snow was more stable than in 1931.

Unlike the Austro–Germans, N.D. Sherpa and Prem Chand were in an ideal position to reach the summit. The snow was in reasonable condition for climbing, perhaps because the ridge had been exposed to westerly winds.

The night of 30 May was not conducive to cooking or sleeping, with the tent hammered by 100-kilometre-per-hour winds. Next morning, 'N.D.' and Chand left the tent at 5 a.m., without breakfast or water since their matches had become too damp to light. They staggered around in the wind, moving to the right of the north ridge, its west side, to try to avoid the wind but had to move back to the crest after floundering in knee-deep snow. The pair climbed on windblasted snow, then on bare rock at a reasonable gradient. They continued over rock, braced against the winds until reaching the western snowfield which stretched unbroken to the summit; they reached it just before 3 p.m. They fixed an aluminium snow stake at their high point, just below the summit, hoping it might be visible through the Himalayan Mountain Institute's powerful telescope in Darjeeling. Fourteen flags, symbolic of a young nation finding its feet, were hoisted. After forty minutes on the summit, they started down, taking a tumble on the way back to Camp 7, which they reached at 8.20 p.m.

The whole party arrived back in Lachen village on 9 June, just three months after the advance party had left to set up Base Camp. The expedition had been highly successful, particularly the summit push. The only downside was the poor performance by the local porters and the Sherpas. Because of that, a lot of equipment and stores were left either on or at the base of the mountain.

It had been a difficult, tightly budgeted expedition for Narinder Kumar, in spite of army support. They could not afford much in the way of the latest western equipment and lightweight food. But they had succeeded where two crack Austro–Bavarian groups had failed. A few years later, a successful Indo–Japanese ascent returned with photographs of the snow stake and flags two metres below the summit. As Kumar declared on his return, 'We never trampled the peak.'

Of the three possible routes leading directly to the summit, as indicated by Freshfield, two had now been climbed after observing and overcoming the dangers Freshfield had warned of. The third way was finally followed two years later.

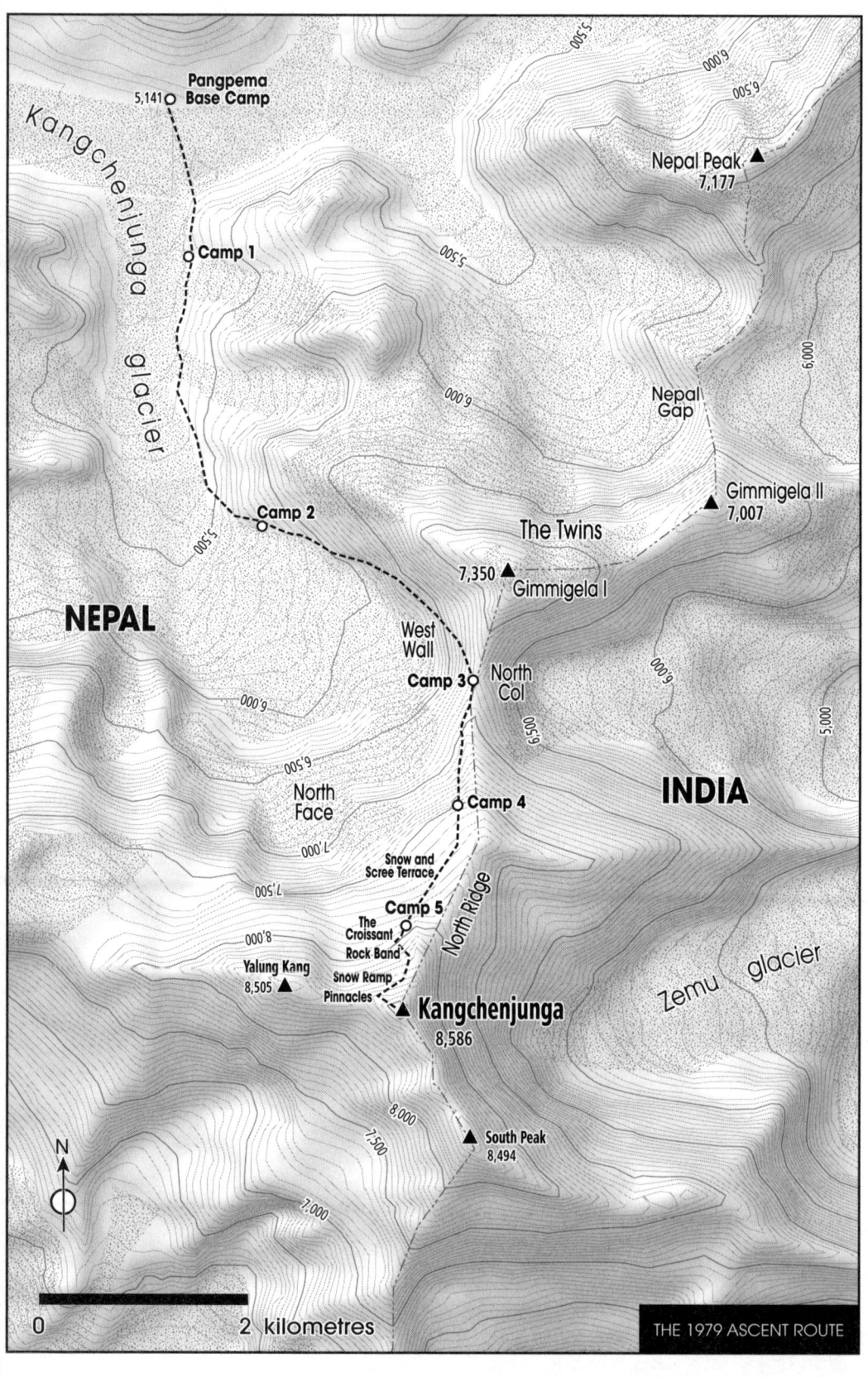

16

The First Ascent without Supplementary Oxygen

1979: Scott, Boardman, Tasker, Bettembourg; Sirdar: Ang Phurba

Kangchenjunga had always meant a lot to me, since I heard as a fifteen-year-old from Gordon Mansell at White Hall Adventure Centre about the fabulous Joe Brown dropping hammers.[1] He had completed the first ascent without a belay! Brown, at only twenty-four, crag climbing at 8,500m ...

On 12 March I walked along the windy, wet street with Martha, taking her to school for the last time for a while and wondering what she really thought of me leaving yet again.[2] I wished she could come, too. At home, Joe Tasker and I weighed our gear: it appeared to be well over the 300 kilos that Air India had allowed us but, fortunately, their representative in Delhi was an Indian mountaineer, Joginder Singh. Within two and a half hours of arrival, we and all our kit were on our next flight to Kathmandu. We had learnt vital lessons: never lose contact with your baggage and never allow it to enter the black hole of a customs shed.

Kangchenjunga had only been climbed by six people, on two expeditions. Both expeditions had used oxygen. We were planning an ambitious new route by the north ridge, as a team of four without oxygen. Joe Tasker and I constituted fifty per cent of the climbing team; Peter Boardman and Georges Bettembourg, the other half. Georges had crossed our path the year

---

1   White Hall Adventure Centre is between Buxton and Whaley Bridge, Derbyshire. It was the first local authority outdoor education centre; courses began in 1951. Gordon Mansell was an instructor there; Scott climbed with him on Arran as early as 1957. 'Dropping his hammer' is probably an allusion to Brown's first attempt, in 1948, at *Cenotaph Corner* (5.10+), which ended at ninety feet when he dropped his piton hammer (actually a mason's hammer) and flattened his belayer.
'Crag' is a reference to Joe Brown's technical climb, without a belay, through a rock outcrop immediately below Kangchenjunga's summit, on the first ascent.
2   Martha: Scott's second child, b.1973, from his first marriage, to Janice Brook.

before, while en route to K2. A Frenchman living in America, he had been enlisted on the strength of his impressively fast ascent of Broad Peak. He arrived at our Kathmandu hotel early next morning, full of ideas for running up hills to aid acclimatisation during the walk-in. Joe, still recovering from a hectic week amassing gear while trying to run his climbing equipment shop, was less than receptive. Leaving Joe in bed, I ran up the road in the cool of the morning with Georges, past the Yellow Pagoda. We then set off to meet Mike Cheney at the Sherpa Co-operative, to attend to everything needed for departure in four days' time. There, we were introduced to our beaming liaison officer, Mohan Bahadur Thapa, a lieutenant in the Nepali army.

We spent the rest of the afternoon in Mike Cheney's office. I was a little miffed to find that unscheduled extras had crept in, mainly two mail runners requiring identical kit to the high-altitude Sherpas, even down to heavy double boots hardly likely to help their speed!

Kathmandu was far busier than on my first visit in 1972. One day, we were nearly run over by an overland trekking lorry which turned out to be driven by 'Jimmy', who had begun climbing with my own Nottingham club. We negotiated a lift to Dharan for all our gear. The day finished with a curry and too many beers at the British Embassy club. I fell asleep and woke up next morning at the hotel: Georges had carried me home.

It was less easy to leave on Saturday than I had hoped. Mike was upset with us hiring a British driver, since such jobs should go to the Nepalis. And Pertemba was upset by our haggling over the price of a bottle of medical oxygen. I had a word with P.T. He said, 'No problem. You want low price, I want high price. It's business, sahib.'

I did empathise with Mike's aims, through his Co-operative, to improve the lot of the Sherpas and porters of Nepal, but at times like this he stretched old loyalties a long way. At least his heart was in the right place.

At last, we could fly to Dharan, where the Gurkhas were allowing us to use their GHQ as a stopover and mail-delivery post. Behind the whitewash and jacaranda, we met our sirdar, Ang Phurba, and sorted our gear into forty-eight porter loads. I had first met Ang Phurba on Everest in 1975 and knew we would be in the best of hands.

The walk-in

Two days later, our route led us along the Leotil Kholi Kosi (river) and across a bridge made in Westminster, before climbing to a ridge we would be following for several days. Walking along its crest between the Arun and the Tamur, the four of us were getting along well, with a good track and a breeze

THE FIRST ASCENT WITHOUT SUPPLEMENTARY OXYGEN

from the west, the sun shining all morning, and occasional views of the Everest–Makalu groups in one direction and Kangchenjunga in another. It did not seem to matter that the wind was blowing a plume off Everest, while the same wind was blowing clouds off Kangchenjunga's summit. I took comfort in the belief that this remorseless westerly would be deflected by the monsoon build-up towards the end of May, the conclusion we had reached from our reading, particularly of the Indian ascent two years before.

We continued north-easterly along this great arm of high ground. The path must have been used for thousands of years; it was such a natural route: the soils are thinner and the vegetation not so dense above what would once have been jungle. Gradually, I began to let energies flow: it was like being plugged into a ley line. There could hardly be a place more conducive to clear thinking, with occasional glimpses of 'Kangch' – the shimmering symbol, neither good nor bad, not seen in relation to anything else, just there, in the 'moment of being'.

We passed Nepalis, some struggling under enormous loads. They seemed to be from a different planet. I asked Ang Phurba why it was that Nike, the Sherpa foreman, did not carry a load: was he really necessary?

'Well,' replied Ang Phurba, 'do you want to lose one load or two, or do you want to pay £18 for Nike's wages?'

'Enough said.'

Our diet included a lot more dhal and beans than usual. Over breakfast, we were asked what the English was for back blast (bean-induced farting): 'backulance' was our reply. Good-humoured banter prevailed. Pete, who had avoided the runs, was goading Joe by asking him if he was ever going to shake them off. Joe kept asking Pete what he was writing. 'Is it another *Boring Mountain*?' – a reference to Pete's first and prize-winning book, *The Shining Mountain*. For hours we gently prodded each other's weaknesses into the open. Nothing went unnoticed. I tried to keep a little detached from the ego games so that I could save something of the calm this ridge engendered for the mountain, where it would be needed.

I was reading the teachings of Gurdjieff in Ouspensky's *In Search of the Miraculous* and growing increasingly aware of my negative tendencies. I confided to my diary:

> On the morning of the 24th I wake up so fully aware of them, they come like a kick in the stomach, a slap across the face, and I find them too embarrassing to write down. My tendency to impress, my overwhelming need to be liked and the consequence that the

opposite is usually the case – it's like giving myself a beating, this holy war I have declared upon myself as I 'fight with my twin, that enemy within'.

Later, I surfaced from a deep sleep to thoughts of Rosie, chuckling away, turning to her mother for milk with a growl of happy anticipation.[3] I again saw Martha, skipping along to school, clutching her reading book. I lay back and wallowed in their loveliness.

We descended from our ridge to the heat of the valleys. A feature of Nepal is the spread of the pipal trees which have survived the slash and burn on the hillsides. Goats and sheep will not eat them. You find them where trails cross or at the centre of a hamlet where travellers can rest their loads in the shade.

One valley appeared richer than most, having substantial two-storey houses with tiled roofs. Our path took us across a rickety chain bridge whose hand-forged links, wearing thin with the passage of people and animals, made one acutely aware of the turbulent stream fifteen metres below. We followed a herdsman who had just purchased a cow for 400 rupees. Ang Phurba declared 'That is £16.' We joked about the cow: it would be cheaper to take it up to Base Camp than pay Kathmandu prices for powdered milk.

Over breakfast one day, we got into a discussion about books, Pete asserting that Heinrich Harrer had overwritten in his New Guinea book *Back to the Stone Age*, and that *Seven Years in Tibet* was too much about himself.

'Oh, come on,' I replied, 'great books in the context of their times.'

Pete reacted rather sharply: 'Don't "come on" me, Dad.'

I think he felt defensive because he knew he had gone too far, but there were also pointers there: I had to guard against pontificating. There were sensitive egos about, and mine was just as sensitive as the others. It did occur to me much later that Harrer's companion, Peter Aufschnaiter, comes across as a shadowy background figure, yet his part must have been at least equal to Harrer's. I was reminded that it is better to listen than to speak.

Our way led uphill towards Ghunsa, the last permanent settlement before Kangchenjunga, a pleasant walk, with a lot of up and down and crossing of little wooden bridges.

Once settled in camp we enjoyed a swim before bouldering: Pete, Georges, Joe and me, in descending order of competence. I enjoyed feeling

---

3   Rosie: Scott's youngest child, b.1978, by his first marriage.

my arms being used for the first time in weeks. There was some harmless-seeming horseplay between Pete and Georges. Georges went up an awkward slab; he was sitting on top and as Pete was about to mount the shelf to join him, Georges gave him a playful push with his foot; down Pete slid, landing with one foot on a tussock of grass, which he slid off with a loud snapping noise. At the least, he had sprained an ankle, maybe even broken a small bone. We supported him to the kitchen fire, and, as Brits do in emergencies, had a cup of tea. Pete and Georges became very depressed: Pete because of the pain, which might take weeks to get better, while Georges was consumed with grief at having caused the injury.

That afternoon, we ate some magic mushrooms; everyone seemed to find them very revealing. I walked a short distance into the forest's green, vibrant interior and sat on a log to contemplate a jungle where each plant was reaching out for the sun to survive. The trees are tall and do not leaf until the top: they form a canopy which does not allow much sun to penetrate it. Some of the vegetation creeps up the trunks in a stranglehold, while the plants on the bottom spread out their leaves to catch what little sunlight does filter through. Using this image to describe some forms of human behaviour seemed apt: I saw our team, point-scoring, vying for each other's attention, for power over the others.

I walked back to find Georges on an incredible trip: he was seeing breasts and penises in everything. Mohan, our liaison officer, could not stop laughing, which helped to take Pete's mind off his troubles.

Transporting Pete remained a major problem. His foot was looking very swollen and multicoloured. First, we tried him on crutches, but that was too painful, so we had a porter make a wicker basket and hired three of them to carry Pete's eighty-five kilos along the trail. I did not think they would succeed but, one step at a time and frequently changing carriers, they managed.

Fortunately, the path was stable. Quite a caravan, including the local relief constabulary and their families, was moving to Ghunsa along the lovely Tamur Valley. We found ourselves happy in the company of a very handsome Sherpani, aged fifty-six, who said she had had twelve children; she told us most of them were now at school in Darjeeling. She was accompanied by two daughters, the elder of whom, Dawa, a pretty girl of about eighteen, became known as 'Smile'. Smile's younger sister, aged about twelve, carried a heavy load. A baby of eighteen months attached to the Sherpani's back completed the group; this could not have been comfortable, as the Sherpani had been kicked by a pig the day before; I had given her some sleeping tablets to help relieve the pain. This family ambled along slowly, exactly what Pete was doing: looking back down the valley to Smile's face gave him the encouragement to put up with his predicament.

Though the path remained well made, the land was no longer rich. A few grim hovels stood high on these grazing uplands from where women came running down to sell gourds full of *rakshi* (local alcoholic spirit).

Pete's porters made him a Mark II carrier and negotiated double wages. We could not argue with that: each was about a third of Pete's weight. Pete himself was getting quite despondent, wondering if he should be returning. Georges was now exuberant then quite melancholic. Pete never really let him off the hook.

Our caravan was smoothly run by Ang Phurba, Kami our cook and our cook's boy, Nima Tamang. All I had to do was sort out the menus. One day, I persuaded Kami to make us nettle soup after I spotted a ragged old lady picking nettles further down the trail. He objected that they were picked in times of famine, so I reminded him that that could be us if he kept giving away bowls of dhal, tea and sugar to anyone who brought us firewood.

Opposite my tent I could see the Sherpani and her four children huddling against the rain under one of our tarpaulins. She was preparing the evening meal. She had gathered wet firewood to get the smoky fire going. She fed her children and put them to bed, all in an hour or so, without any fuss. Everything she did, she did with such grace and economy of effort. I marvelled at that as I sat looking out from my tent doorway, surrounded by tapes, tape deck, books and camera equipment. It struck me just how much of my time I spent doing too many things badly, whereas these people just seemed to carry out life's basic necessities so well: every thought and action seemed to have meaning.

Next day, the wind was blowing in gusts, with occasional rain squalls which helped to keep down some of the black soot from the fire-devastated hillsides. I asked Mohan to ask the local people why there was such extensive damage. It seems a Japanese expedition returning from Jannu in 1974 with 200 porters had set fire to the hillside by accident; it did not stop burning for months. By the time the monsoon put it out, ten houses had been destroyed as well as fifty to sixty cattle. Secondary growth was now hiding the worst of it, but we could still see huge burnt stumps everywhere, for about eleven kilometres.

By next morning, the scene had changed: the rain had given way to snow, which melted as we climbed the zigzag path towards Ghunsa, past grazing yaks and thickets of scarlet rhododendron, with snow peaks above.

The valley opened to a wide terrace, where the trees had been cut back for potato fields. There were wooden houses, some on stilts, and a monastery. We had hoped to be given the abbot's blessing, but he was asleep, so we carried on to Ghunsa, a pleasant place occupying the wide floor of the valley, with thick forests all around. We walked through the village to its far end and started to pitch camp. Before we could get the tents up, dozens of people were queueing for medicines: to help their children take milk; for fat-bellied people who had worms; for those with goitres. They appeared not to be true Sherpas but folk recently arrived from Tibet. We did what we could then went to sleep; another seven centimetres of snow fell.

I was starting to have vivid dreams, as I normally do at altitude: we were now at 3,650m. In one dream I saw myself with a shaven head, walking along a path. I was asked my name and said it was Melinder. Next day, by chance, I found there was a King Milinda's book of Buddhist scriptures.[4] I got quite excited at the coincidence but had to concentrate on immediate jobs such as registering at the police station. Ten policemen, all dejected lowlanders, looked very miserable up there.

We decided on a two-day stay at Ghunsa, to let Pete's foot recover, to acclimatise, engage fresh porters and buy provisions including, at Ang Phurba's insistence, firewood. He said they could not exist without it for two months at Base Camp while we were climbing. Fortunately, there was a lot of dead wood around.

On the walk-in, apart from dipping into Buddhist scriptures and Gurdjieff, I had also been reading Carlos Castaneda's *Tales of Power*. I found it comforting that all three writings agreed on how to achieve real happiness, by emptying the mind, checking internal dialogue and overcoming negative tendencies. Nonetheless, the vast extent of the task before us remained daunting.

On 2 April, we left Ghunsa with our new porters. They were mostly Sherpanis, driving yaks with our forty-seven loads up to Base Camp at Pangpema. An easy walk led to Kambachen (sometimes 'Kangbachen'), a collection of mean huts used by yak herders only in the hot season. Pete walked with us, testing his foot and sharing our delight at the incredible views of Jannu's west face. Kangchenjunga, when we glimpsed it above the Ramthang glacier, did not seem so inaccessible any more. So many other tempting peaks appeared: how good it would be to come up here on a simple trekking permit! The peaks reminded me of the Hindu Kush: dry below, with snow on the top as if it had been spread over the summits with

---

4   King Milinda: Probably a reference to *Milinda Pañha* (lit. 'Questions of Milinda'), a Buddhist text from between 100 BC and AD 200.

an icing knife. The Sherpanis, joined by some of the young Sherpas, danced all the next afternoon and into the evening, beating time on our boxes and barrels – it was all we could do to stand up, never mind dance, having been overtaken by the altitude.

One of the Sherpanis approached to demand rupees. Was it for the privilege of watching them dance? No: she had come up the hillside the day before to put offerings under a Buddhist shrine on a prominent rock.

Base Camp

Our Base Camp site, Pangpema, was a wide terrace of frozen grass, falling very steeply for some 120 metres to the ever-shrinking glacier. But what a place! I have never felt so enclosed by so many big mountains. We were not the first outsiders since the 1930s to visit this side of the mountain: Japanese rubbish was spread all around. According to the locals at Lhonak they had stayed here only two weeks ago on a recce of our route for the following year. I found the remains of a hut – perhaps from the international expedition of 1930 – and some very old wooden tent pegs. Maybe Frank Smythe had slept here.

We retained six porters to assist us up to Camp 1, then helped Kami erect the kitchen tarp over the ancient hut and pitched our tents around it, some fifty spacious metres apart.

One of the porters told us that 300 yaks graze from here all the way to Kambachen in the summer. He told us how he had to leave Tibet ten years ago when the Chinese razed his home. Ang Phurba was haggling with a man who had brought a hundred kilos more wood than we needed. We offered him half his price, to which he agreed, rather than take it all back again. For us, it was a bit of a waste, since we had enough kerosene to do all the cooking. 'But that's not the point,' Ang Phurba argued: the wood was for Kami, Nima Tamang and Mohan to keep warm at night when we were on the hill.

I decided to sleep under a flysheet, pitched over little clumps of flowers. I arranged my mat and sleeping bag and sat outside my tent: opposite us was Kangchenjunga and our route, with the mighty Kangchenjunga glacier squeezing its way between the flanks of the Twins, Ramthang Peak and Wedge Peak, with its fluted snowfields and faces. We had arrived with the minimum of fuss after the best walk-in ever, for which achievement most credit was due to Ang Phurba.

THE FIRST ASCENT WITHOUT SUPPLEMENTARY OXYGEN

Acclimatisation

Next morning, with our six porters, we descended to the Kangchenjunga glacier. It was a mean day, cold and windy, with quite a lot of snow in the air. Pete soon went back: it was all too much for his leg. After five hours plodding up the glacier in thick mist, the rest of us deposited a pile of gear and food just before some séracs, opposite the Twins. Over the next three days we established this as Camp 1. It was marvellous to get back to Base Camp to cook, shelter, drink pints of tea and have Sherpa steamed potatoes lathered in yak butter with a fresh, ground-chilli-paste dip.

On 9 April, after another very cold night (-18 °C), we left late to reconnoitre a way up to Camp 2, going slowly to accommodate Pete, and without Georges, who was feeling ill. We made good progress through the séracs where the upper Kangchenjunga glacier turns east round the base of the Twins. There, close to the steep west face of Kangchenjunga leading up to the North Col, we saw our route for the first time. It looked horrendous, especially head-on. After winding through a second group of séracs, we found a reasonable place for a camp: from here, we could look across the now open valley to the place where Sherpa Chettan was killed on the 1930 expedition. It did seem a foolhardy line for trying to break through Kangchenjunga's defences. I wondered how we would manage on the big wall up to the North Col. We retraced our steps to Camp 1, with Pete in tow, saying he now knew how his guided clients felt. Next day we returned to Base Camp to complete our acclimatisation, a process which cannot be rushed. Pete, who had been suffering from diarrhoea, was despondent and grumpy. Joe disappeared, except for meals.

By Friday 13th, Ang Phurba, Nima and I had returned to Camp 2, to dump our loads, mainly of food. It was a steamy day. We retreated to Camp 1 for endless brews. Georges and I shared a tent next to Pete and Joe. Ang Phurba and Nima camped further down. We seemed to be well protected by two moraine banks from an avalanche chute on the Twins. At 9.30 p.m., we heard an avalanche louder and nearer than the rest. I thought of Graves and *Goodbye to All That*: 'You know by the sound when they have you in mind.' The dust settled; then another came even closer. Georges peeped out of our tent and Pete stood outside his: the ice blocks which had broken off stopped seventeen metres short. We had Smythe's *The Kangchenjunga Adventure* with us: he had observed the same evening crescendo and put it down to meltwater expanding when it cools, thus wedging the ice off.

The four of us next day moved up to Camp 2, just over three kilometres in three hours, with 600 metres of ascent to 5,800m, all part of the acclimatisation process – one day at a time. The day after that, we took a closer look at the north wall, the first big test along our chosen route.

It was a baking hot day. I was coughing painfully from the stomach and became completely drained. We went down the ropes as fast as possible, to rehydrate in the evening, but by then the water was icy and I drank too much. I crawled the last bit into camp. Georges took off my boots and fed me sips of hot water, which was all I could manage before vomiting violently. A humbling lesson: it was one of the hardest days I had ever done in the hills and I felt so stupid at having let myself get so dehydrated.

I spent the next day drinking and drinking again, as Joe and Georges made good progress: just visible with the naked eye, two dots on the ice about a kilometre away, approaching 6,700m. At least ten big avalanches fell around us while big black crows pecked at our rubbish.

On 19 April, Pete and I started up the ropes put out by Georges and Joe the day before: I was impressed by what they had done. It was hard work even getting up them, especially one slightly overhanging pitch which Georges had led. I led up another 150m, pulling over short overhanging rock barriers of granite, crabbing across hard, brittle ice on crampon points and ice picks with no thought beyond the next step, totally absorbed in finding a way to our North Col goal. I moved slowly to a snow ramp to a short block step and poured myself over that with my last reserves of energy. Now, it seemed that only open slopes remained to the North Col. But it was already 6 p.m. We reached camp by 7.15 p.m. Joe and Georges seemed pleased with our progress.

The mail arrived suddenly the next day, with Ang Phurba and Nima; we all took ourselves off to immerse ourselves in it.

Georges and Joe set off first on 21 April to reascend the ropes and complete the route to the col. Pete and I tramped up the glacier half an hour behind, and struggled up the ropes with tents, gas, food and sleeping bags, all weighing about twenty kilos. I felt I was gathering energy as the four of us reached our high point. Pete trailed in last, but he too was on the mend.

The usual flurries of snow in the afternoon did not go away; they only got worse: suddenly, we were in a full-scale blizzard. There was only a short rock step and angled snow barring the way to the North Col, but we had to go down, retreating in an epic of avalanche powder and the threat of bigger stuff pouring down from the flanks.

After a night at Camp 2, we reached Base Camp by noon for a meal of egg and chips, wonderful after ten days on the mountain. It snowed all afternoon and into the night. By now, we had run out of rope for fixing and used up all our climbing ropes as well: at home, Joe had calculated the rope in feet instead of metres, so we had only a third of what we needed. He was suitably chastened when he realised what had happened. Our only chance of acquiring more rope lay with a Czech expedition on Jannu, so we sent Ang Phurba down with half a gallon of whisky and crossed our fingers. Meanwhile, we continued to enjoy the luxuries of base-camp life, such as macaroni cheese and fresh baked bread and honey. Past exploits dominated the evening conversation, all of us joining in with well-known, oft-repeated stories.

We spent another rest day and to our amazement, that evening, Ang Phurba arrived with 150 metres of seven-millimetre Czechoslovakian nylon rope, a tin of very strong plum spirit, which we drank with stewed apple fried in butter, and the offer of their medical facilities should we need them.

Then we were off to Camp 2 again, with more food and the rope, all set for a 4 a.m. start. It is depressing how things can change overnight. Georges had not slept and Joe was suffering from headaches. But Pete was well, having slept without sleeping pills. It was incredibly cold outside: -18 °C, so we lay where we were until the sun came up. Reading *The Way of Zen* did nothing to help me rationalise my situation. I feared that time was slipping by and that I would never be home before June. Next morning, we tried again.

The first attempt
Georges and I led off, carrying about eleven kilos each, with our ski poles for higher up. We reached our high point by 2 p.m. Pete and Joe were not in sight, so Georges and I loaded up with an enormous weight of camera gear, stove, pans, two days' food, down suits, over-suits, shovels, spare clothes, spare glasses, goggles, candles, head torches – it was all I could do to stand up, never mind pull myself up the ropes at 6,700m. Georges took off, leading out for forty metres. I led for another fifteen metres to easier slopes. Georges then led through a chimney of rotten rock.

Unfortunately, he knocked off some rocks while leading and I knocked off more while hauling my gear behind me. As Georges headed off, Joe appeared from below to shout through the now stormy weather that Nima had psyched out lower down and Ang Phurba had gone back with him; also that Pete had been hit on the hand by one of the falling rocks: he thought it might be broken. I could see him start to make a painful descent with Ang Phurba and Nima.

Everything that had started so well that morning was now falling apart. Still, Joe and I plodded on after Georges to just below the col. The most spectacular views of Everest and Makalu appeared through a slot in the seething golden clouds above the valleys, the big mountains black against mauve and purple skies.

Eventually, we collapsed into a little tunnel tent which Georges had brought from Seattle. We were absolutely done in after this monumental effort. It was very cramped, especially for Joe, who lay at the entrance, over the slope. Georges groaned all night, with the wind howling outside, but the tent stood up well considering it weighed only two kilos. I slept fitfully, trying to keep myself from rolling on to others.

At 7 a.m. I left the tent to relieve myself. The only windless spot I could find was over the col in Sikkim. And that was how we made it to the North Col of Kangchenjunga. There was complete calm on the other side, with brilliant early morning views eastwards into the rising sun coming up behind Siniolchu.

We migrated there straightaway, digging out a platform and half a snow cave to accommodate one of our two tents. We moved into the tents as the snow fell and, looking down the Twins glacier, wondered how far up John and Joy Hunt had reached when exploring a way to the North Col just before the war.

Joe was now suffering excruciating headaches, for which he was taking large doses of the strong analgesic fortral. He had arranged that Ang Phurba would bring up a load that day and let us know about Pete's damaged hand. The advantage of a small group like ours was that everyone knew what to do without awaiting direction, like Joe when he oversaw Nima and Pete's descent.

From Castaneda, I noted:

> A warrior cannot complain or regret anything, his life is an endless challenge and challenges cannot possibly be good or bad ...
> Without the awareness of death, everything is ordinary, trivial ...

I could not disagree, for up here it made enormous sense. But I knew that down below it was a different matter.

That afternoon, thunder was in the air – something I had never really noticed before when high up. We could all imagine the massive falls of snow to follow. My tent was looking more spartan now, as bit by bit I had shed my burden, leaving behind everything inessential. Once, a book to read and a book to write in might be considered superfluous. I read a bit more from the teachings of Don Juan and noted:

THE FIRST ASCENT WITHOUT SUPPLEMENTARY OXYGEN

> The best of us always comes out ... when we feel the sword dangling overhead. Personally, I wouldn't have it any other way.

And neither would I, now that at last we seemed to be en route and committed, despite our problems: Joe was flat out with a steamy headache and there was still no word of Pete from Ang Phurba.

They arrived the next day; Pete was able to climb. Sadly for Joe, another day and night of pain and nausea spelt descent for him.

On 1 May, it was agreed that Pete, Georges and I would continue to the ridge to check it out and leave some food as high as possible, then descend to sleep at the col. Joe's problems made us very aware that we had to monitor our own acclimatisation very carefully. Georges was in his element, moving fast for two and a half hours, winding his way up snow and ice between pinnacles of rock until we stopped for a lunch of kippers and tea, then continued for another hour to the foot of a tower of crumbly rock. There seemed to be a good opportunity here to make a snow cave in a bank of wind-blown snow. To the left, the slope led up 100m to the big rock band below the great scree terrace.

It took us only an hour to descend to the col – we were pleased about that, but sad to see Joe, a tiny dot below, walking along the track into Camp 2. We lay around under heavy snowfall, discussing what to do next. Georges was all for going for it, Pete for a cautious build-up and protecting our retreat. None of us knew how fit we would be up there. But we were eager to try for the summit and there was no talk of waiting for Joe.

I had a wonderful seven-hour sleep and woke, fully refreshed, to begin the morning rituals. We stuffed our sleeping bags and some spare clothing into our sacks, added a tent, left a note for Joe and Ang Phurba, and set off up the ridge in a harsh, cold wind. Georges and I followed Pete, unroped. This cheered him up: he was obviously on the mend. When we reached our little cache we began to dig a cave and took stock of our situation. There were no signs of people around Camp 2. It looked as if Joe had gone all the way down to Base Camp. There was not a cloud anywhere and the view was clear to all horizons: being on a ridge was far more enjoyable than the claustrophobic south-west face of Everest.

We eventually dug through both sides of the snowbank and, after meeting in the middle, began to hollow out the interior. At 7,470m, it was hard work and took us three and a half hours. But we did a good job and ended up with a huge, safe haven against storms. We had no energy left to check out the rock steps above, so spent the afternoon melting snow for brews and filling our water bottles for the morning, full of optimism that we could make our bid for the top next day – then be on our way to Kathmandu and home to our families.

That night I did not sleep well. I had earache and took two aspirins, the first pills I had taken on a mountain since 1972. I seemed to be getting another cold and blamed myself for getting so wet by digging the cave without my over-suit on. Throughout the night, Georges fidgeted in his half-sleep, got up to look for a mug to pee in – at which Pete and I protested! Despite all this, we got away by 7 a.m. I led up soft, deep snow for about 150m to the rock step. We left the last of the Czech rope hanging below us to secure our retreat.

The climbing was quite difficult, up granular snow with no place for solid anchors but, as we were in a shallow couloir, we were sheltered by the rocks on our right. I pulled up and over to the great scree terrace and met the full force of the westerly winds. I could hardly stand: time to go down to our ice palace for kippers and brews. We discussed what to take with us the next day and felt guilty about Joe. If we did not leave soon, we would have to go down for more food.

By 10 a.m. we had reached the screes of the great terrace. Pete was going well but Georges was very tired: I seemed to be towing him across the terrace, five paces then a pull on the rope, usually mid-step, a real drag. The wind was vicious, strong enough to blow flaky bits of rock uphill. During the worst gust, we had to huddle together until a slight lull allowed us to jump up, stumbling and tripping over the boulders until the next blast had us cowering, backs to the wind, clutching our down and windproof suits. At least we were cocooned against the cold. I tried to rationalise the wind's impact by reasoning: 'It's only air particles, only wind moving around the globe – get with it, lad, you're not freezing yet.'

Then we would resume, staggering on another few paces.

The nearer we got to the ridge, the stronger and noisier the wind became. On the crest, we became the first obstruction the westerlies had met since Everest. I felt an electric charge in my back, like a hundred needles jabbing into my spine, addling my marrow. The only way out lay down, through the static of hundreds of flying particles, to the relative calm of Sikkim.

About twenty metres down we started the laborious, three-hour job of hacking a tent platform out of the steepening slope that dropped 2,000m to the Zemu glacier. Enough room, but not enough oxygen at 7,900m with the door shut and the fumes from the stove. I struggled to melt ice for tea and, somewhat more grudgingly, for a hot-water bottle for Georges' cold feet.

THE FIRST ASCENT WITHOUT SUPPLEMENTARY OXYGEN

I took a headache pill and slept well, until Pete woke me at 1.30 a.m. to help with our wind-hammered tent: the central fibreglass hoop had broken. By 2.30 a.m., the wind was extreme, as was our position. We panicked our way into over-suits, boots and gaiters, working in shifts as Pete and I took turns to hold the broken ends of the hoop to prevent it tearing the fabric. Our gloved hands were wet and cold. Then, every climber's nightmare: the whole tent was blown a metre along the ledge, towards the drop. I tore the zip open and bounded outside, heels dug into the ledge, gripping the tent, trying to prevent it, Pete and Georges from plunging down to the glacier.

This was the wildest wind I had ever known: chips of ice flew in my face and blasted my nylon suit. I grabbed an ice axe and rammed it through the tent fabric where the ties had ripped off. I stamped it down, but the wind simply ripped the tent away another few centimetres, leaving me holding on with both hands, as poles snapped and bits of fabric tore off. Georges yelled for his crampons; I flung them in, yelling to pack everything in our rucksacks: 'Don't abandon anything!' I was determined that this was not going to be a rout. The outer sheet of the tent stripped off and flew, pale orange, into the cold night. And all the time I clung on, wrestling with the surviving hoop, finally twisting it down and breaking it to lower the tent's resistance. I felt I was losing, with my frozen hands and failing strength, scared I was going to lose the gear, my friends and myself, all tumbling down to the Zemu ...

'Surely, this isn't the way I'm going down – blown off a bloody mountain, sliding into an unknown valley – come on, get a grip.'

I shouted for them to get out quick. First a Karrimat came through the entrance, followed by rucksacks then Pete. Georges was last, cutting his way out of the back with his Swiss penknife, still drowsy from too many sleepers. As he emerged, we let go of the tent and it flew off, another black shape fast disappearing. It became a dot in the distance, then was gone.

It was now 4.30 a.m. We squatted on that icy ledge, two metres long and just over a metre wide, in the strongest wind any of us had ever experienced. Summit thoughts were non-existent; indeed, we still had to get ourselves down.

For the first hour we clung to each other, moving through a stupor of cold and utter misery. Pete was not saying anything; he seemed to be getting hypothermia. I thought that this time I had gone a step too far. For the second time in my life, I thought I was going to die.

Our situation was deteriorating fast. Georges realised this first and, in a burst of energy, dashed off up the ropes we had tied off to a deadman belay plate three metres above us. From this, he made a bold attempt to reach the col twelve metres away, but the rope was wind-tangled and pulled him up short. He attempted twice more before reversing to our ledge. My turn. As I stood up, I felt so unstable that I dipped back down, my head bunched into my chest, hiding from the flying ice and snow. I wrenched my sack off, grabbed two ice axes and pulled up to the deadman plate. Then, with frozen fingers, bit by bit, I untangled the flailing rope. Georges joined me and paid out the rope as I reached a point just below the col. I came to a dead halt, blasted by the winds, crouching on crampon points and ice picks, my face pressed against the ice, trying to find some relief from the turbulence.

Georges shouted, 'Doug, go for eet, go for eet!' but I missed a slight lull when I could have risked moving. 'Go for eet, Doug!' yelled Georges again. I went for it, scampering up on blunt crampons, scraping and scratching at the hard green ice in a continuous flailing of limbs, trying to reach the top before the next onslaught. I threw myself over the ridge and squirmed above Nepal on my stomach, literally clawing my way down the slope with ice picks to some rocks, where I tied off.

Without goggles and glasses, my face was covered in ice, with a coating of frozen snot and sputum on my beard and hair, but I was now out of immediate danger. I thought about getting my hair cut at a certain barber's in Kathmandu – the first alien thought after hours of total concentration on survival.

Eventually, Georges squirmed over the ledge.

'Doug, my sack, eet has down the Zemu glacier!'

'Don't worry, kid, we'll get it all together!' I yelled, as we both pulled Pete over to join us. We began to stagger and shuffle down to the cave bivouac at 7,620m, and arrived there at 10 a.m. We had climbed down to the col by 1 p.m. then continued down the north face to Base Camp two hours later. Joe could not stop talking. After all the worry of waiting, he fussed around us, plying us with brews and food. We had descended 3,000m, and covered eight kilometres of the mountain all in a day, but then we had got off to an early start!

This first attempt at Kangchenjunga had been my closest call. I will never forget it, because I was oblivious to all else but surviving the wind and the cold when, at George's insistence, I 'went for eet', to face the unknown with adrenaline running high, exhilarated, detached, free and climbing beyond ego, going for it.

During the next few days, I lay on the grass under my flysheet, dreamlike, watching animated scenes of home behind my eyelids. I wondered if I had learnt anything. I just felt so much at peace, so relaxed with myself and my friends. I would always feel warmth towards them from having shared that ordeal, and to Joe, who had contained his disappointment at not being with us, but who then took care of us.

I sat on the grass at Base Camp, my bottom bony and my body skinny, nursing my frostbitten finger-ends, stinging from strong antibiotic cream, wrapped up in Joe's white silk gloves. Although I now had excuses to leave, there was Kangchenjunga, still smoking snow and mist. What a mountain it was turning out to be! I now knew many of its secret places, at least to 8,000m on the north-western flank, but I did not know about that last 600m.

We took rest days on 7 and 8 May. By 9 May, all of us had returned to Camp 2. Reaching it only took some three and a half hours from Base Camp, showing that we must have recovered.

During the next two days we planned to climb to camps 3 and 4, then take off from the snow cave for the summit, climbing all night if all went well with the weather, to get ourselves down during daylight hours before suffering too much from the altitude. There was not a breath of wind and, from appearances, none at the top either: just our luck! My frostbitten fingers were mending, but blisters had appeared on my right hand's small and ring fingers, like burns or scalds. Pete had a frost blister on his big toe, and a numb nose and ear. Georges had recovered from his snow-blindness and Joe was raring to go.

The second attempt

Next day, we put the plan into operation, reaching Camp 3 on the North Col at noon. After digging out our tents, we settled in for an early start in the morning. Joe set off first, so he could choose his pace. It was very tiring in the wind, with relatively heavy packs. Returning to a slope is never as enjoyable as the first time, and the third time definitely becomes boring. Joe and I arrived at the snow cave to find the roof was sagging. I spent an hour digging it out, followed by three brews and a bowl of soup. Although the blisters had gone down, my frostbitten hands were beginning to ache in the wet cold. We were all tired and talked about taking a rest, in the hope of giving the wind time to settle and for the moon to light our way.

I resisted asking Georges for a hot-water bottle. The summit of Kangch was only a day or two away now, after all this effort. About twelve gas cylinders and enough food for a week meant we should be in with a chance of rounding this face of the mountain to the summit ridge, to see Joe Brown's crag and all the land to the east. Georges did the cooking and even produced a hot-water bottle for me after all.

I woke at 6.30 a.m. on 12 May to wish Joe a happy thirty-first birthday, then sat outside the cave reading more of Don Juan. We needed this rest day to gather our resources for the next push. Each of us was locked into his private apprehension about going 1,200m higher at these altitudes with no artificial oxygen, with more than 600m over unknown ground. As if in warning, wind and snow flurries kept us locked in our little prison with the sagging roof. We stayed put, making brews to relieve our tongues, which felt like strips of leather.

We snoozed and brewed the next morning away. I had to push away thoughts of home, coming to the only conclusion possible, that I just had to see what else this mountain had to show me. 'Maybe one day I will grow out of all of this,' I thought, but for the present I knew that was why I was there. Clouds came scudding in from the west, which frightened me but, to the east, Tibet was a delight.

We enjoyed a final brew just before it grew dark, then prepared to set off up the ropes, with Georges in front pulling them out of the frosted snow. Spindrift was blowing into our faces. We were pleased at our fast pace: by 8 p.m., we had arrived on the scree terrace, but we worried about the clouds which kept enveloping us, weakening the moon's light to a limpid glow and obliging us to use our head torches to pick out a rock cliff we had called the 'Croissant'. We stayed on one rope and had to keep stopping for Pete to adjust his loose crampons. As midnight approached, with the gusts and snow flurries blowing harder, we stopped for a brew beneath a huge boulder and thought about where to spend the night. Just below the Croissant we came to a suitable place for a cave. At least, Joe and I thought it was suitable, and we started to dig it out. Pete and Georges decided to push on to nearer the rocks to look for a better place. They never came back to report on their probings, so we continued to dig our shelter and settled in, shivering, for the night, sipping hot water since Pete had taken the Rise and Shine with him!

It grew so cold that I resorted to my old trick of digging out more of the cave just to create warmth. After the windiest of nights, we emerged at 3.30 a.m. to the wildest of mornings. We attempted to join Georges and Pete, but the wind was far too violent. We only just found our way back to our own cave. An hour or so later we emerged again to see Georges standing about 150m above us; we battled our way to him. Georges shouted into the wind, 'We have just come down from an attempt.' I cursed him.

'We are supposed to be a team, and you eff off with both ropes, most of the gear, not saying anything, in one of the worst snow and wind storms, in the middle of the night on unknown ground and on the third-highest peak in the world!'

Georges looked away, crestfallen, into the hole in the bergschrund where the two of them had spent the remains of the night. Pete looked up at me, his face pleading for understanding. 'I just got carried away with Georges' momentum,' he added, rather lamely. We all knew how impulsive Georges could be once he was in 'going for eet' mode. They both looked so dejected that I apologised for my outburst. Everyone was shattered, especially Pete and Georges. They had in fact gone about another 180m to 240m higher on sloping soft snow at forty degrees before retreating to their cave.

We were all quite done in after a night in such primitive conditions, with only the clothing we had walked up in. Without the storm, we might have made it to the summit, but now we had to retreat again to thaw out in our big snow cave at Camp 4. We reached it at 9 a.m., dug out wind-blown snow from the entrance and got the stove on. We then sat around in our sleeping bags, pulling frozen snot, mucus and water vapour out of our beards, taking stock of the situation between brews. We all agreed that we should not have gone in these weather conditions.

Pete and Joe were keen to have another go, but Georges was definitely going down. I admired the others' lack of ambivalence, especially Georges, and the strength of his conviction that he had burnt himself out having, as he said, psyched himself up for the big push with no reserves remaining. In any case, he did not believe the weather would improve. I kept my thoughts to myself as we talked in general terms about what we were doing up there, just the four of us without the logistical support of a big expedition. What if one of us got pulmonary oedema? How would the others get him down? And it might be me, we were all thinking.

This first day back at the cave produced a clear head, returned strength, warmth and safety. I felt quite calm, but for how long? I knew myself too well by now: soon, sinking thoughts would creep in. The pull of home would only get stronger if Georges wanted to go down. With these thoughts, others followed:

'Isn't it time for me to get out of Himalayan climbing? Shouldn't I be leaving it to the younger lads like Pete and Joe? What if Georges was right and the weather never improved? Could we spend another two weeks on the mountain? Ang Phurba reckoned there was going to be a lot of snow before the monsoon came in. Another two weeks without getting anywhere, with Pete and Joe who will give nothing away, so withdrawn into each other. Had I the strength to cope if I stayed – or if I quit and they succeeded?'

Georges felt that we had, in effect, 'done' Kangchenjunga, and it had nothing left to show us. But I knew that was not true. Going to the Himalaya has always concerned having to cope with altitude. It was not like Mont Blanc, where doing a hard route on its flanks is satisfying enough without having to plod up to the summit. Of course, the journey is more important than arrival, but this journey is not over until the summit is reached. With these thoughts I started to feel stronger. There was no way I could give up now.

I tried to persuade Georges that the four of us could make it, but he was not having any of it. We would all have to descend, as there was not enough food or fuel to continue from here to the summit. Down at Camp 3 on the North Col we could restock and gain strength for one more go.

I woke, totally refreshed, at three in the morning from a very solid six hours' sleep. I pushed the snow blocks away from the entrance to our cave and looked outside to a beautiful starry sky and, for the first time, not a breath of wind roared through the rocks above. At five o'clock I put on a brew, woke the others and told them I had strong feelings that we should go straight back up now. Pete gave a guarded 'yes' and then, while shuffling around on his Karrimat, discovered two full gas cartridges; he thought that was a good omen. Joe was still feeling very tired and hesitant but could see the logic in it – snatching good days whenever they came. I told him how powerful my instinct was – as if I had had a visitation. Joe, a former Jesuit seminarist who had put all that behind him, told me exactly what I could do with that notion! We rummaged around the cave and gathered half-used packets of nuts, dates, bars of chocolate, soup – just enough to make a lightning two-day dash for the summit. Joe said he would give it a try. I turned to Georges, but he declined, repeating that he had no reserves left. He reaffirmed his conviction that he had already been as close as he needed to be to the summit, that he had reached the pinnacles on the summit ridge. Pete's face showed disbelief, but he said nothing. We started packing: Georges with all his personal gear and we, with our sleeping bags, two gas cylinders, one stove, two billies, camera, film, two Karrimats each and a minimum amount of first aid. We wore only the clothes we stood up in; our only extra gear was one rope.

We emerged into a beautiful morning and watched Georges head off down. Joe started towards the terraced rock band. He gave up, exhausted, after a few steps in the deep snow. I went ahead while the wind picked up. Pete became dubious about our chances. We agreed to continue to the top of the ropes and make a decision on the scree terrace. By that time, the wind had died down. I shouted down to the others that there was cloud on both sides. 'They seem to have neutralised each other!' I shouted. After belaying the others, I left them resting and set off across the great scree slope, building cairns on prominent boulders as I went, just in case the clouds did roll in and we found ourselves in another storm. I wondered as I built the highest cairned footpath in the world what the anti-cairn men in Scotland would have to say about that!

I located the hole that Pete and Georges had sheltered in and found a considerable bergschrund grotto against the rock of the Croissant. I hacked at it for an hour until Pete arrived. He then joined me; an hour after that, Joe came in. I cooked and brewed until 9.30 p.m., to fill our water bottles. My finger-ends were really hurting: it felt as if the frostbite was getting deeper. A cold wind started to blow through the cave. Pete climbed back up the shaft to seal off the top before we finally settled in for the night. I took half a Mogadon – just for the hands, I told myself – although really I was disappointed that I had succumbed to this chemical.

Pete was up first next morning. He climbed straight up the shaft and shouted down exuberantly that there were blue skies and not much wind. After setting about the slow ritual of filling bottles and having more brews, Pete led off up the thirty-five-degree snow slope, in good condition. Joe and I then went ahead: from below, with Georges' binoculars, we had scrutinised this part of the summit cone for weaknesses. An obvious way lay up to the right, round the upper part of the Croissant. Joe led through a short rock step. I brought Pete up, and Joe led again, kicking his way through deep snow.

I was getting worried about my feet: the new Messner inners were a bit cramped inside my old Makalu boots and I was afraid I was going to get frostbitten toes as well, so I stopped to drink some water and take my boots off to rub some life back into my toes. It was sunny and warm. While I was doing this, Pete led on past Joe and into a gully directly below the black pinnacles. To the side of this gully was a jet-black rib of basalt. Pete reached the summit ridge and looked down at us, grinning, saying that he could see all the way down the west and south ridges to Jannu and Kabru. We soon joined him. It was wonderful to look over the other side after being confined to the north-west face for so long. Pete and Joe were obviously feeling quite pleased with themselves, especially Pete who, against all the odds, had coped with his injuries and had finally taken a turn in leading the way. Behind him, huge, billowy clouds reached up hundreds of feet. Jannu looked very small below us and quite insignificant; Kabru was free of cloud: we could plainly see the route the Norwegians had taken in 1907, a great flat area as big as a football pitch. After taking photographs, we turned to the summit, agreeing to go on until we felt the gods would be angry with us. I led off. Although there was no double vision, no hallucinating, I was not going very fast. Even plodding on a few more steps was not that easy. I had to scramble over rocks, crawl over parts of the ridge on all fours and lower myself down a gully to reach the east side of the summit ridge. Joe Brown and George Band had come up somewhere around here. We passed under the famous *Joe Brown's Crack*. I headed for a large jutting flake of granite. Under it, I found a natural cave which I registered as a possible bivvy *in extremis*. But when I stood on top of the flake, I realised we were almost at the summit. Joe and then Pete slowly joined me. We sat in the snow, then walked up the last few feet, to stop three metres from the summit. This last section had been a hard grind. We regained our strength; I tried to circumnavigate the summit, to see all sides of the mountain without having to go over the top. It would not be easy and, in any case, Pete and Joe were chivvying to go down. I took some more photographs and they chivvied me more. There was no trace of the British or Indians.

Pete remembered he needed some photographs of Karrimor products, so I took some more shots and we all relaxed. I began to feel quite euphoric – more than I ever did on top of Everest.

It was so good to be on top with just a butty bag on our backs, without, for me and Pete, the oxygen bottles we had to carry on Everest. These were just not worth the weight and the trouble, not if we thoroughly acclimatised first; that was what we had achieved by going high on two occasions, coming down to rest, and then finding we were good for 8,586m on the third push. What a day it had been: still no wind, while above all those clouds an enquiring *gorak* flew higher and higher, higher than Everest.[5]

After a final sip of water, we retreated. We had reached the summit area at 5 p.m. and remained for about three quarters of an hour, so we were descending in twilight on very blunt crampons, into a golden sunset over the western approaches. It was a slow descent in light snow and lightning; I found that my tinted glasses did not help my vision, especially after the lightning, which seemed to darken them. I stumbled over one rib of snow on the great diagonal traverse but luckily stopped in a pile of soft stuff underneath. Pete took over the lead, as he was obviously moving faster than I; he led us straight to our cave by nine o'clock.

The lightning was still flickering as we spotted lights at Base Camp. They

---

5   *Gorak*: probably onomatopoeic (perhaps from Greek κοραχ); a Himalayan bird closely related to the Alpine chough.

too could see us on the upper Kangchenjunga cone! We got into our sleeping bags, slightly apprehensively, as we had hoped to have reached the big cave at 7,470m. Other than feeling a bit hungry, though, we enjoyed a good night's sleep.

Making it to Camp 4 next morning, we picked up all our remaining gear, then on we went down the north ridge in good weather to the col, where we repacked tents and clothing and made up a big bag to throw down from the col. The next section was hazardous from loose rock, as so much snow had melted since our upward passage: we sank up to our thighs. Nima and Ang Phurba came up to meet us. After another night's sleep we soon regained Base Camp.

As I walked up the final rocky slopes, Georges came down to greet us. We flung our arms around each other. He seemed so genuinely pleased that we had climbed our mountain. I told him that there had been a lot of him up there with us. He was certainly there in spirit. We all flopped out on the grass and started drinking tea, eating egg and chips, followed by a long pancake session into the night. It felt so good to be alive, safe and well.

We had shown ourselves that it was possible to climb one of the big mountains of the world without major sponsorship and its attendant media hype. With a bit of help from our Barclays Bank friend, Alan Tritton, and the Mount Everest Foundation, we had managed to cover all the expenses with a contribution of only £1,600 each. That, of course, would not have been possible had we opted to take oxygen bottles, as had every other expedition that had been to the Three Big Mountains of the World.[6]

Consequently, we were able to climb Kangchenjunga in lightweight fashion, without the need for an army of porters to Base Camp, and anything up to a hundred Sherpas operating along fixed ropes. There was no need for winches or radios, as there is with a big expedition with its inevitable command hierarchy. Nor did we need extra personnel to take care of the equipment or the Sherpas or radios, or to man base and advance base camps.

This first-hand experience of climbing light and high was going to be useful for our future confidence as Himalayan climbers. The real significance of our experience on Kangchenjunga was far more personal. I came away full of admiration for Pete and Joe. Pete had the strength to keep going, and to take part in lead climbing on the summit day. Joe, who had more or less resigned himself to going no higher than 7,000m, had also kept going until

---

6 Oxygen: Messner and Habeler had of course climbed Everest without supplementary oxygen in 1978, 'as their own team'. They were, however, part of a larger German team led by Wolfgang Nairz, which used oxygen.

his strength returned, then overcame a severe throat infection to lead a difficult rock section at 8,230m. They had both shown astounding reserves of mental and physical stamina.

We were at times in great danger: then, we became so close to each other that we became one animal. I am still amazed when I look back to the hurricane episode, at how we found the energy to survive. We had centred ourselves on the act of survival; any stray thought had been banished from our minds. There was no anger, no envy, no frustration, not even anxiety; we just bent our energy to one end – to survive and get down. There was also the camaraderie, for such crises make you very close: you see each other as you really are; energy flows, one to the other. For a time, we all radiated the understanding and awareness that are normally hidden from each other because of our egotism.

I will always enjoy the day I spent with Georges when we first climbed to the North Col in fine weather, up solid granite, with excellent snow and ice. We led without our sacks, wearing only minimal clothing. I delighted in Georges' exuberance, free of all restrictions, moving fast over mixed ground.[7]

And of course, I take great delight recalling my time spent with Pete and Joe on summit day, picking our way up the mountain in the hope that we were doing the right thing. All three of us had left for Kangchenjunga full of ambition, but within that upper region no thought occupied our minds other than just being there. All that was left was the childish delight of living among the clouds, concentrating on the next step, and seeing where it would take us. As luck would have it, one step at a time, we had gone all the way to a few steps from the summit.

---

7   Georges Bettembourg died, aged thirty-two, on 18 August 1983 from stonefall on the Aiguille Verte.

The Kangchenjunga Massif: Treks, Attempts and Ascents

1848–50	Sir Joseph Hooker: two scientific treks as far as the Kangbachen Valley.
1854–56	Hermann Schlagintweit: meteorological survey on the Singalila Ridge.
1861	Major James L. Sherwill, Captain E. MacPherson, Dr B. Simpson, W. Kemble: tourist trek to Guicha La area.
1876	Elizabeth Mazuchelli with husband Francis and 'C': tourist trek; cross the Kang La to Jannu.
1879–81	Pundits Sarat Chandra Das ('S.C.D.') and Ugyen Gyatso ('U.G.') cross the Chorten Nyima La en route to Tashi Lhunpo; Maurice de Déchy and Andreas Maurer reach the Singalila Ridge.
1883	William Woodman Graham makes a disputed ascent of Kabru.
1883	Pundit Rinzin Namgyal ('R.N.') explores the Talung Valley.
1884–85	Rinzin Namgyal ('R.N.'): first clockwise circuit of the massif.
1899	Douglas Freshfield with Rinzin Namgyal ('R.N.'), Vittorio and Erminio Sella, Edmund Garwood, Angelo Maquignaz, Erminio Botta: anti-clockwise circuit of the massif.
1905	Aleister Crowley, Dr Jules Jacot-Guillarmod, Alexis Pache, Alcesti C. Rigo de Righi, Charles Reymond: first attempt at climbing Kangchenjunga.
1907	Norwegians Carl Rubenson and Ingvald Monrad Aas reach $c.7{,}300$m on Kabru.
1907–21	Alexander Kellas's six expeditions.
1920	Harold Raeburn and Colonel H.W. Tobin, then Raeburn and C.G. Crawford: expeditions to southern aspects of the massif.
1925	Nikolas Tombazi: photography expedition to the Guicha La and Talung glacier area.

1926	Captain J.E.H. Boustead's disputed crossing of the Zemu Gap.
1929	Edgar Francis Farmer's disappearance near the Talung Saddle.
1929	Paul Bauer's first expedition: the north-east ridge.
1930	Günter Oskar Dyhrenfurth's International Expedition; first ascent of Jonsong Peak.
1931	Paul Bauer's second expedition to the north-east spur of Kangchenjunga.
1935	Reginald Cooke and Gustav Schoberth ascend Kabru.
1936	Paul Bauer's third expedition: Siniolchu ascent.
1936	Bill Tilman's first expedition: Zemu Gap area.
1936	Marco Pallis in the Zemu glacier area.
1937	Reginald Cooke, John and Joy Hunt visit the Zemu glacier and nearly reach the North Col.
1937	First German–Swiss expedition: Ludwig Schmaderer, Herbert Paidar and Dr Ernst Grob in the Zemu area.
1938	Bill Tilman's second expedition, while returning from Everest; crosses Zemu Gap from the north.
1938–39	Ernst Schäfer's *Ahnenerbe* expedition to Sikkim and Tibet.
1939	Second German–Swiss expedition (Ludwig Schmaderer, Herbert Paidar and Dr Ernst Grob); ascent of Tent Peak.
1942	Group-Captain A.J.M. Smythe and Wing-Commander L.S. Ford climb Lama Anden.
1945	Chomiomo climbed by Harry Tilly, Ang Tharkay and Dawa Thondup.
1945	Wilf Noyce climbs Pauhunri.
1949	Swiss expedition led by René Dittert to the Kangchenjunga massif: numerous ascents.
1951	George Frey and Gilmour Lewis: reconnaissance of south-west face.
1953	Gilmour Lewis and John W.R. Kempe: reconnaissance of the Yalung face.
1954	John Kempe, Gilmour Lewis and Trevor Braham appraise the Yalung face.
1955	First ascent led by Charles Evans. Summiteers: George Band and Joe Brown, then Norman Hardie and Tony Streather.
1976	Indian army training-selection expedition on Siniolchu.
1977	Indian army ascent via the north-east spur led by Colonel Narinder Kumar. Summiteers: Prem Chand and Nima Dorje Sherpa.
1979	First ascent of Kangchenjunga without supplementary oxygen. Summiteers: Doug Scott, Pete Boardman and Joe Tasker.

Select Bibliography

Books

Aris, Michael: *Bhutan: The Early History of a Himalayan Kingdom*, Aris & Phillips Ltd, Westminster, England, 1979

Bauer, Paul (translated by Hall, E.G.): *Himalayan Quest*, Nicholson & Watson, London, 1938

Baume, Louis C.: *Sivalaya: Explorations of the 8000-metre peaks of the Himalaya*, Gastons-West Col Publications, Reading, Berkshire, 1978

Bernbaum, Edwin: *Sacred Mountains of the World*, University of California Press, Berkeley, California, USA, 1998

Bista, Dor Bahadur: *Fatalism and Development: Nepal's Struggle for Modernization*, Orient Blackswan, Hyderabad, India, 1991

Boardman, Pete: *The Shining Mountain*, Hodder & Stoughton, London, 1978

Braham, Trevor: *Himalayan Odyssey*, George Allen & Unwin, London, 1974

Bruce, C.G.: *Himalayan Wanderer*, Maclehose & Co, London, 1934

Buchanan, Francis: *An Account of the Kingdom of Nepal and the Territories Annexed to this Dominion by the House of Gorkha*, Constable, Edinburgh, 1819

Burrard, Sir Sidney: *A Sketch of the Geography and Geology of the Himalaya Mountains and Tibet*, Superintendent, Government Printing, India, 1908

Cameron, Ian, in association with the Royal Geographical Society: *Mountains of the Gods*, Century, London, 1984

Castaneda, Carlos: *Tales of Power*, Simon & Schuster, New York, 1974

Cooke, C. Reginald: *Dust and Snow: Half a Lifetime in India*, C.R. Cooke (privately published), 1988

d'Anville, Jean-Baptiste: *Nouvel Atlas de la Chine, de la Tartarie Chinoise et du Thibet*, The Hague, 1737

Desmond, Ray: *Sir Joseph Dalton Hooker: Traveller and Plant Collector*, Antique Collectors' Club, Majestic Books, London, 1999

Duff, Andrew: *Sikkim: Requiem for a Himalayan Kingdom*, Birlinn Limited, Edinburgh, 2015

Du Halde, Jean-Baptiste: *Description géographique, historique, chronologique, politique et physique de l'Empire de la Chine et de la Tartarie Chinoise*, Paris, 1735

Evans, Charles: *Kangchenjunga: the Untrodden Peak*, The Travel Book Club, London, 1956

Filippi, Filippo de (Ed.): *An Account of Tibet: The Travels of Ippolito Desideri of Pistoia, S.J., 1712–1727, with an introduction by C. Wessels*, George Routledge & Sons, London, 1937

Fowler, Mick: *Vertical Pleasure*, Hodder & Stoughton, London, 1995

Frawley, David: *Gods, Sages and Kings: Vedic Secrets of Ancient Civilization*, Lotus Press, Twin Lakes, Wisconsin, USA, 2012

Freshfield, Douglas: *Round Kangchenjunga: a Narrative of Mountain Travel and Exploration*, Edwin Arnold, Publisher to HM India Office, London, 1903

Gansser, Augusto and Olschak, Blanche Christine: *Himalayas*, Facts on File, St Louis, MO, USA, 1988

Goodman, Martin: *Suffer and Survive: gas attacks, miners' canaries, spacesuits and the bends – the extreme life of Dr J.S. Haldane*, Simon & Schuster, London, 2007

Graham, W.W., Thomson, J. and Markham, A.H.: *From the Equator to the Pole: Adventures of recent discovery by eminent travellers*, W. Isbister, London, 1887

Gregson, Jonathan: *Blood Against the Snows*, HarperCollins, India, 2002

Hagen, Toni: *Nepal: the Kingdom in the Himalayas*, Kümmerly & Frey, Bern, 1960

Hale, Christopher: *Himmler's Crusade: the true story of the 1938 Nazi expedition into Tibet*, Bantam, London, 2003

Hardie, Norman: *On My Own Two Feet: the life of a mountaineer*, Canterbury University Press, Christchurch, New Zealand, 2006

Harrer, Heinrich (translated by Graves, Richard): *Seven Years in Tibet*, Rupert Hart-Davis, London, 1953

Hedin, Sven: *Southern Tibet: Discoveries in the former times compared with my own researches in 1906–1908*, Lithographic Institute of the General Staff of the Swedish Army, Stockholm, 1917–1922

Hooker, Sir Joseph: *Himalayan Journals*, John Murray, London, 1854

Hunt, Sir John: *Life is Meeting*, Hodder & Stoughton, London, 1978

Hunter, Sir William Wilson: *Life of Brian Houghton Hodgson, British Resident at the Court of Nepal*, John Murray, London, 1896

Isserman, Maurice and Weaver, Stewart: *Fallen Giants: a history of Himalayan mountaineering from the age of empire to the age of extremes*, Yale University Press, New Haven and London, 2008

Kapadia, Harish: *Into the Untravelled Himalaya*, Indus Publishing, New Delhi, India, 2005

Keay, John: *The Great Arc: the dramatic tale of how India was mapped and Everest was named*, HarperCollins, London, 2000

Kielkowsi, Jan: *Kangchenjunga Himal and Kumbhakarna Himal*, Explo Publishers, Poland, 1999

Kirkpatrick, William: *An Account of the Kingdom of Nepaul, being the substance of observations made during a mission to that country in the year 1793*, W. Muller, London, 1811

Longstaff, Tom: *This My Voyage*, John Murray, London, 1950

Lunn, Arnold: *A Century of Mountaineering, 1857–1957*, George Allen & Unwin, London, 1957

MacGregor, John: *Tibet: A Chronicle of Exploration*, Routledge & Kegan Paul, London, 1970

Madge, Tim: *The Last Hero: Bill Tilman*, Hodder & Stoughton, London, 1995

Majupuria, Trilok Chandra and Majpuria, Indra: *Sacred and Useful Plants and Trees of Nepal*, Tribhuvan University, Kathmandu, Nepal, 1978

Manandhar, Narayan P.: *Plants and People of Nepal*, Timber Press, Oregon, USA, 2002

Markham, Clements R.: *Narratives of the Mission of George Bogle to Tibet and of the Journey of Thomas Manning to Lhasa*, Trübner, London, 1876

Mason, Kenneth: *Abode of Snow: a history of Himalayan exploration and mountaineering from earliest times to the ascent of Everest*, Rupert Hart-Davis, London, 1955

Mazuchelli, Elizabeth Sarah (writing as 'A Lady Pioneer'): *The Indian Alps and How We Crossed Them*, London, 1876

Mitchell, Ian and Rodway, George: *Prelude to Everest*, Luath Press, Edinburgh, 2011

Morgan, Gerald: *Ney Elias: explorer and envoy extraordinary in high Asia*, George Allen & Unwin, London, 1971

North, Marianne: *A Vision of Eden: the life and work of Marianne North*, Webb & Bower in collaboration with the Royal Botanic Gardens in Kew, Exeter, 1980

North, Marianne (edited by Symonds, Mrs J.A.): *Recollections of a Happy Life*, Macmillan & Co, London, 1892

Olschak, Blanche Christine & Gansser, Augusto: *Himalayas*, Facts on File, St Louis, MO, USA, 1988

Ouspensky, P.D.: *In Search of the Miraculous: fragments of an unknown teaching*, Routledge & Kegan Paul, London, 1950

Patey, Tom: *One Man's Mountains: essays and verses*, Victor Gollancz, London, 1971

Peissel, Michel: *The Ants' Gold: the discovery of the Greek El Dorado in the Himalayas*, HarperCollins, London, 1984

Pierse, Simon: *Kangchenjunga: Imaging a Himalayan Mountain*, University of Wales School of the Art, 2005

Roerich, Nicholas: *Altai-Himalaya: a travel diary*, Frederick A. Stokes, New York, 1929

Roerich, Nicholas: *Himalayas: Abode of Light*, London: David Marlowe, 1947

Roerich, Nicholas: *Shambhala*, Frederick A. Stokes, New York, 1930

Searle, Mike: *Colliding Continents: a geological exploration of the Himalaya and Karakoram of Tibet*, OUP, Oxford, 2017

Sella, Vittorio (with Ansel Adams, Greg Child, Paul Kallmes and Wendy M. Watson): *Summit: Vittorio Sella – Pioneer mountaineering photographer 1879 to 1909*, Aperture, New York, 1999

Shaha, Rishikesh: *Ancient and Medieval Nepal*, Manohar Press, New Delhi, 1997

Shor, Thomas K. and Palmo, Tenzin: *A Step Away from Paradise: The true story of a Tibetan lama's journey to a land of immortality*, Penguin India, New Delhi, 2011

Singh, Mahendra Man: *Forever Incomplete: The Story of Nepal*, SAGE Publications, New Delhi, 2013

Smythe, Frank S.: *The Kangchenjunga Adventure*, Victor Gollancz, London, 1930

Smythe, Frank S.: *The Spirit of the Hills*, Hodder & Stoughton, London, 1935

Smythe, Tony: *My Father, Frank: Unresting Spirit of Everest*, Bâton Wicks, Sheffield, UK, 2013

Subba, J.R.: *Mythology of the People of Sikkim*, Gyan Books, New Delhi, India, 2009

Symonds, John: *The Great Beast: the life of Aleister Crowley*, Ryder & Co. London, 1951

Tilman, H.W.: *When Men and Mountains Meet*, CUP, Cambridge, 1946

Tombazi, Nikolas: *Account of a Photographic Expedition to the Southern Glaciers of Kangchenjunga in the Sikkim Himalaya*, Bombay, 1925

Tucker, John: *Kangchenjunga*, Elek Books, London, 1955

Twain, Mark: *Following the Equator*, The American Publishing Company, Hartford, Connecticut, USA, 1897

Unsworth, Walt: *Hold the Heights: the Foundations of Mountaineering*, Hodder & Stoughton, London, 1993

Vajracharya, Dhanavajra and Malla, K.P.: *Gopalarajavamsavali*, Nepal Research Centre, Kathmandu, Nepal, 1985

Waddell, Major Laurence A.: *Among the Himalayas*, J.B. Lippincott Company, Philadelphia, USA, 1899

Wangchuk, Pema and Zulca, Mita: *Khangchendzonga: Sacred Summit*, Hillside Press, Kathmandu, Nepal, 2007

Wessels, Father Cornelius: *Early Jesuit Travellers in Central Asia, 1603–1721*, Martinus Nijhoff, The Hague, 1924

Yule, Colonel Henry: *Cathay and the Way Thither*, Hakluyt Society, London, 1866

Guidebooks

Reynolds, Kev: *Kangchenjunga: A Trekkers' Guide*, Cicerone Press, 1999

Wheeler, Tony & Everist, Richard: *Lonely Planet: Nepal*, Lonely Planet Publications, 1990

Journals, periodicals and newspapers

Alpine Journal
Climbers' Club Journal
Englishman, The
Geographical Journal
Himalayan Journal
India Pioneer
Journal of Applied Physiology
Journal of the Asiatic Society of Bengal
Mountain
Pioneer Mail, Allahabad
Proceedings of the Royal Geographical Society
Rock and Ice Club Journal

Websites

www.sacredtexts.com

Index

A
Aas, Ingvald Monrad   126, 130–1, 147, 149
Afghanistan   17, 51
Ahnenerbe expedition   177
Akademischer Alpenklub Berlin   182
Allwein, Eugen   179, 190, 192–5
Alpine Club   85, 89, 92, 93, 95, 95n., 103, 110, 123, 127, 131, 136–8, 152, 155, 157, 158n., 163, 164, 215, 216, 221, 221n., 223
   German and Austrian Alpine Club (DÖAV)   194, 195, 196n.
   Italian Alpine Club   98
Alpine Journal   89, 93, 94, 107, 110, 123, 126, 127, 129, 130, 137, 152, 154, 155, 156, 158, 159, 159n., 164, 193, 201
American Alpine Journal   165, 167n.
Ang Phurba, Sherpa   233–6, 238–45, 251, 255
Ang Thari, Sherpa   171
Ang Tharkay, Sherpa   171, 208
Ang Tsering, Sherpa   171, 201
Animist religion   19, 30
Annullu, Sherpa   224, 226
Araniko   26
Arun (river and valley)   vii, 8, 27, 57, 67, 234
Arunachal Pradesh   30, 202n.
Ashoka   17–18, 20
Asiatic Society of Bengal, Journal of the   50, 106
Assam   7, 22, 47, 202, 207
Aufschnaiter, Peter   190, 192, 194, 195, 201, 236
avalanches   12, 13, 95, 120, 125, 135, 143, 144, 147, 162, 163, 166, 170, 185, 186, 187, 188, 190, 194, 198, 199, 213, 216, 217, 225, 226, 230, 241, 242

B
Bagmati   25, 26
bamboo   30, 64, 67, 68, 74, 75, 79, 104, 113, 159, 160

Band, George   222, 225–7, 254
Bangladesh   17
Base Camp
   German–Swiss   200, 201
   Kabru   171
   Kangchenjunga   94, 117, 182, 183, 185, 236, 239–43, 245, 248–9, 254–5
   Lhonak   210
   Norwegian   148
   Yalung   140–2, 166–7, 224–6
   Zemu (Green Lake)   173–4, 177, 182, 189–91, 193, 195, 198–9, 204, 229
Bauer, Paul   179, 181–2, 189–92, 194–9, 202, 204, 213, 229
Beger, Bruno   203–5
Beigel, Dr Ernst   190, 192–3, 194
Benares (Varanasi)   43
Bengal   17, 23, 27, 35, 40–2, 49–52, 81, 90, 120, 163, 172
Bengal, Bay of   8, 56, 113
Bettembourg, Georges   233–4, 236–8, 241–53, 255–6, 256n.
beyul   viii, 30, 30n.
Bhaktapur   24
bharal (blue sheep)   60n., 72, 161, 204
Bhutan   17, 28, 36, 39, 40, 41, 42, 51, 207
Bhutia   viii, 28, 29, 30, 70, 104, 173
Bhutia Boarding School   51, 58, 59
birch   15
Blaser, Willy   123, 128
Boardman, Peter   233, 235–9, 241–56
bodhi (pipal) tree   17, 236
Bogle, George   41, 41n., 42, 52
Bön   19, 19n., 38n.
Boss, Emil   124, 127, 128, 129, 131, 132, 133
Bourne, Samuel   95
Boustead, J.E.H.   165, 178
Braham, Trevor   208, 209, 215, 216

Brahmaputra 8, 50
Brenner, Julius 179, 190, 192, 194
bridges 52, 53, 78, 97, 113, 119, 160, 176, 186, 234, 236
Brown, Joe 223, 225, 226, 227, 233, 233n., 249, 254
Bruce, General 116, 127, 157
Buchanan, Francis 43, 44
Buddha 22, 49
Buddhism vii, 17n., 18, 19, 20, 22, 23, 24, 24n., 25, 26, 30, 49, 51, 55, 58, 85, 91, 104, 179, 229, 239, 239n., 240
bungalows 65, 140, 173, 204, 208, 216
butterflies 113, 119, 225

C

Cabral, Father João 36
Cacella, Father Estêvão 36
Cairngorm mountains 151, 152n.
Calcutta (Kolkata) 45, 47, 48, 50, 51, 52, 71, 96, 106, 109, 112, 120, 124, 149, 170, 172, 190, 205, 208
Campbell, Dr Archibald 49, 65, 69, 75–7, 81–2, 84
Campbell, Robin 145, 146
Castaneda, Carlos 239, 244
cedar, Himalayan 14
Chabuk (Chubuk) La and glacier 56, 60, 61, 117, 210, 212
Chand, Colonel Prem 229–30
Changabang 130
Chapman, Freddy Spencer 173, 179, 199, 211
Cheney, Mike 234
chestnut 113, 159
Chettan, Sherpa 184, 186–7, 192, 241
Child, Greg 30
China 17, 18, 22, 26, 39, 40, 51, 202, 207
Chomolhari 39, 41, 67, 81, 173, 199, 208, 228
Chorten Nyima La 55, 56, 61, 117, 155
Chumbi Valley 28, 36, 42, 58, 81, 140
Chumiomo 155
Chunjerma pass 53
Claude White, John 97, 111
climate viii, 11–13, 19, 45, 71, 79, 95, 106
Climber & Rambler 146n.
Collie, Professor Norman 128, 129, 131, 132
Conway, Sir Martin 127, 128, 137, 138
Cooch (Kuch) Behar 36, 39, 40
Cooke, Reginald 129, 169–75, 179, 195, 199, 200, 211, 213
Cooke's Cooker 171
'coolies' (also porters) 52, 54, 55, 57, 66, 67, 78, 91, 99, 105, 111, 114–7, 124, 125, 135, 137, 139, 140, 142–6, 148–50, 152–7, 159–62, 163–6, 168, 169, 171, 173, 183–7, 190, 192, 194, 195, 198, 200, 204, 208, 210, 216, 222, 224, 225, 229, 230, 234, 237–41, 255
Cooper, Julian 93–5
Crawford, C.G. 161–2, 163, 166, 167, 208, 214
Crawford, Captain Charles 43–44, 47
Croissant, the 94, 250, 253
Crowley, Aleister 135–47, 153, 161, 162, 216, 217, 221

D

Dalai Lama 90
d'Anville, Jean–Baptiste 35, 37, 39, 41
Darjeeling (Dorjiling) 8, 9, 11, 27, 44, 47–52, 57–9, 62, 64, 65, 66, 69, 71, 72, 75–9, 82, 83–91, 93, 94, 96, 103–6, 108, 109, 111–13, 116, 119–22, 123–24, 129, 132, 135, 138–40, 144, 146–8, 152, 153, 156, 157, 160, 161, 163, 165, 166, 169–75, 178, 182–4, 188, 194, 195, 198, 207–8, 210, 212–6, 224, 225, 229, 230, 237
Das, Sarat Chandra ('S.C.D.') 51–8, 60, 61, 117
Dawa Tenzing, Sherpa 222, 223, 227
Dawa Thondup, Sherpa 173, 174, 175, 208, 211
Déchy, Maurice de 123
Delhi sultanate 18
Desideri, Ippolito 37–8
Devil Dance 122
Dharan 234
dharma 24, 24n.
Dittert, René 210–12
Domvile, Admiral Sir Barry 202
Donauland 196
Donkia La 50, 57, 79, 80, 81, 84, 153, 195
Donkia Range 28, 81
Du Halde, Jean–Baptiste 35, 37, 39
Dyhrenfurth, Hettie 181, 182, 184
Dyhrenfurth, Professor Dr Günter 168, 181–2, 185–8, 194, 214

E

earthquakes 24, 25–6
East India Company 44, 49, 50, 84, 110
Eckenstein, Oscar 136, 137, 138, 142
Evans, Charles 157, 162, 195, 221–7
Everest, Mount vii, 9, 10n., 45, 48, 67, 85, 91–3, 116, 125, 131–2, 151, 152, 155–8, 161, 162, 165, 170–1, 173, 175, 176, 184, 185, 194, 201, 204, 208, 210, 212, 215, 221n., 222–3, 226–9, 234, 235, 244–6, 254, 255n.

F

Falconer, Hugh 7, 7n.
Farmer, E.F. and Mrs 165–8, 184–185
Farrar, Percy 152, 154, 157
fig (pipal) tree 17, 236

INDEX

Filippi, Filippo de 38
Fitch, Ralph 40
Fitch, Walter 83–84
flora and fauna 13–15, 17, 45, 65, 83, 86, 156, 229
Fluted Peak 92, 179
Forked Peak 93, 130, 148
Freshfield, Douglas 10, 52, 55–7, 62, 77, 98–9, 106–21, 127–8, 139, 153, 157, 162, 181, 185, 189, 190, 212, 231
Frey, George 213–14

G

Ganges and Gangetic plain(s) 7, 8, 11, 17, 19–21, 36, 47, 49, 62, 201
Gangtok 28, 111, 112, 147, 160, 169, 178, 188, 201, 203, 204, 208, 209, 224
Gangway, the 226–7
Garwood, Professor Edmund 111, 114, 121, 128, 139, 141, 146, 185
Geographical Journal 130, 152
German Mountaineering and Hiking Association 196
G(h)unsa 53, 54, 57, 60, 61, 119, 184, 185, 212, 225, 236–9
Ghunsa K(h)ola 10, 53–5
Gimmigela I and II *see* Twins, the
Godwin-Austen, H.H. 48, 82
Goecha (Guicha) La 10, 91, 93, 105, 120, 124, 159, 163–5, 175, 176, 178
Gopala 20, 21
Göttner, Adolf 198, 199, 200
Graham, W.W. 123–33, 135, 147
Grand Plateau (or Shelf) 120, 162, 213, 216, 226
granite 6, 10, 72, 74, 76–7, 189, 200, 242, 254, 256
Great Trigonometrical Survey (GTS) 49, 50, 58, 62, 103, 127
Green Lake 113, 154, 173, 177, 189, 191, 195, 199, 229
Grob, Dr Ernst 173, 174, 200, 201, 212
Guillarmod, Dr Jules Jacot- 135, 138, 139, 142–5, 217
Gupta empire 18
Gurkhas (Ghorkhas) 20, 24, 28, 42, 43, 67, 75, 111, 127, 152, 234
Guru Rinpoche viii, 28
Gyaljen, Sherpa 161–2, 184
Gyatso, Lama Ugyen ('U.G.') 51, 52, 57–61, 117

H

Hahn, Dr Kurt 215n.
Haldane, Professor J. S. 157, 158
Hamilton, Alexander 41, 42
Hannah, J.S. 182, 184

Harman, Captain H.J. 50, 52
Harrer, Heinrich 190, 201, 236
Harrison, Ginette 106
Harrison, Lieutenant J.B. 179, 211
Hastings, Warren 40, 41
Hedin, Sven 37, 202
Hepp, Günther 198, 199
high altitude, physiology of 151, 152, 153, 157, 158
Highlands, Scottish 152, 160
Himalaya
    Eastern Himalaya ix, 6, 11, 19, 22, 27, 64, 99, 106, 224
    foothills vii, 12, 42, 44, 45, 47, 49, 201
    Great Himalaya 6, 8, 62, 103
Himalayan Club 161, 169, 172, 179, 189, 190, 200, 204, 208, 209
Himalayan Journal 9n., 10, 92, 159, 168, 170, 208, 211, 228
Hindu people and culture 5n., 17, 19, 23, 24, 25, 209
Hinduism 17n., 24
Hindu(stani) language 5n., 222
Hodgson, Brian 44–6, 65, 66, 70, 78
Hoerlin, Bettina 196, 197
Hoerlin, Herman(n) 181, 181n., 182, 185, 186, 188, 194, 196, 197
Hoffmann and Johnston 96, 97
Hooker, Sir Joseph 14, 49, 50, 53, 57, 63–82, 83–7, 104–5, 108–9, 119, 121, 122, 126, 140, 156, 169, 181, 212
Hughes, Glyn 123, 128
Humboldt, Alexander von 65, 84
Hunt, Lady Joy 172–5, 213, 244
Hunt, Sir John 157, 172–5, 200, 212–13, 221, 244
Hunter Workman, William 127

I

ice-cave 192
India
    army 104, 209, 213, 228–30
    country vii, ix, 6, 8, 11, 14, 17–20, 22–5, 27, 28, 30, 36–40, 42, 43, 49, 50, 51, 59, 70, 71n., 82, 84, 85, 87, 88, 90, 91, 92, 95, 96, 97, 103, 106, 111, 129, 139, 147, 163, 170, 190, 202, 207, 209, 215, 222, 224, 228
    government 29, 51, 57, 59, 96, 181, 190, 202, 209, 225n.
Indian Plate 6
Indus 7, 18
Inner Line 225, 225n.
International Himalayan Expedition 181–8, 200

J

Jackson, John   215, 223, 225–6
Jackson, Ronald   215
Jacot-Guillarmod, Dr Jules   135, 138, 139, 142–5, 217
Jannu (Kumbhakarna)   10, 27, 53, 60, 74, 89, 94, 99, 107, 109, 119, 120, 132, 161, 238, 239, 243, 253
Japanese   230, 238, 240
Jesuits   35, 36–7, 39, 252
Joe Brown's Crack   254
Joint Committee of Alpine Club and Royal Geographical Society   215, 221, 222
Jongri (Dzongri)   52, 72, 119–21, 148–9, 165, 168, 214
Jonsong (Jongsong) La   9, 55–7, 60–2, 99, 117, 153–4, 183, 188
Jonsong (Jongsong) Peak   56, 92, 117, 154, 155, 188, 194, 210
Journal of the Asiatic Society of Bengal   50, 106
Jubonu   124, 159

K

Kabru   9, 9n., 74, 89, 93, 105, 120, 123–32, 141, 147, 148, 151–3, 156, 164, 169, 170–2, 215, 253
Kabur   94, 120, 129, 148, 165
Kalimpong   9, 82, 112
Kambachen (Kangbachen)   10, 54, 55, 60, 72, 73, 85, 119, 183, 185, 211, 217, 239, 240
Kamet   123, 151, 155, 156, 157, 228, 229
Kang La   9, 53, 57, 72, 74, 91, 107, 119, 124, 132, 141, 151, 156, 165, 168, 182, 183, 184
Kangchenjunga
  Central Summit   9
  circuit   55, 58, 59, 62, 77, 99, 109, 111, 114, 116
  east side   10, 97, 106, 171, 254
  glacier   10, 60, 94, 118, 153, 183, 185, 187, 188, 212, 240, 241
  massif   5, 6, 9, 10, 12, 27, 29, 31, 42, 58, 59, 76, 79, 90, 92, 94, 97, 98, 105, 106, 113, 114, 120, 135, 152, 154, 161, 163, 179, 195, 197, 198, 208, 210, 212, 213, 225
  National Park   14, 69
  North Col   174, 185, 191, 199, 213, 241, 242, 244, 249, 252, 256
  north ridge   99, 113, 174, 185, 186, 191, 194, 213, 230, 233, 255
  north side   51, 114, 161, 210
  south side   72, 77, 191
  South Summit   9, 11, 164
  summit (and ridge)   viii, 9, 10, 31, 48, 77, 88, 106, 114, 119, 120, 122, 135, 141, 158, 185, 187, 192, 211, 217, 224, 226–8, 230, 231, 233n., 235, 245, 249, 251–6
  west side   8, 10, 12, 13, 50, 211, 213, 215, 230
  Yalung Kang (West Summit)   9, 10, 59
Karakoram   6, 42, 127, 137, 138, 173
Kathmandu
  city   20, 22, 24–7, 37, 38, 43–5, 47, 97, 183, 209, 210, 221, 233, 234, 236, 246, 248
  Valley   8, 18, 19, 20, 21, 22, 24–7, 42–5, 97
Kaufmann, Ulrich   124, 125, 129, 131, 133
Kellas, Alexander   131–3, 151–9, 158n., 162, 163, 177, 179, 188, 208, 211, 212
Kempe, J.W.R.   214–7, 221, 225
Ketar (Kitar), Sherpa   186, 193
Kinloch of Kinloch, Captain George   42
Kiranti   6, 20, 21, 22, 27, 69
Kirkpatrick, William   42–4
Kitan Jigmay, Sherpa   171
Klaproch, J.H. von   42
Knox, Captain W.O.   43
Kokthang   9, 74, 124, 214, 215, 225
Kolkata   *see* Calcutta
Kosi   7, 8, 44
Kumar, Colonel Narinder   228–31
Kurz, Marcel   181, 188, 214
Kyirong Valley   25

L

Lachen (place and valley)   61, 62, 79, 80, 81, 111, 119, 154–6, 163, 173, 174, 193, 195, 198, 204, 229, 230
Lall, John   209
Lama Anden (mountain)   156, 208
larch, Himalayan   73
Lear, Edward   83, 85–6, 89, 92
leeches   68, 140, 148, 159
Lepcha (Lapcha)   vii, 28–31, 64, 66, 69–71, 73, 75, 77–8, 96, 99, 104, 108, 112, 156, 160, 175, 195, 203, 209
leucogranites   6
Leupold, Joachim   190, 194, 195
Lewis, Gilmour   213–6
Lhakpa, Sherpa   177, 178
Lhasa   26, 27, 37, 38, 41, 41n., 52, 91, 112, 202, 205
Lhonak (place and glaciers)   55, 56, 117, 118, 154, 179, 185, 188, 210–2, 240
Licchavi people and dynasty   20, 22–3
Limbu   19, 22, 27, 28, 29, 30, 69, 104, 209
Limbuwan   27, 209
Lobsang, Bhotia   184
Lobsang, Sherpa   165–6
Lohner, Annelies   210, 211
Longstaff, Tom   110, 126, 128, 129, 133, 137, 155, 157
low temperatures   15, 170

INDEX

M
magnolia   113, 159
Mahabharata   6, 21, 22
Mahabharata Range   7, 7n., 9
Makalu   vii, 85, 92, 235, 244
Malla dynasty   23–4, 26, 40, 42
Maoist insurgency   26, 27
Maquignaz, Angelo   111, 113
Markham, Sir Clements   37, 41n., 52
Mason, Kenneth   8, 116, 128–31, 170, 189, 190
Mather, Neil   223, 225, 227
Matthews, Donald   215
Mazuchelli, Elizabeth Sarah (Nina)   53, 57, 106–9, 119
McKinnon, Tom   222
Mechi (people and river)   27, 44
migmatites   6
Mirgin La   53, 73, 74, 119, 183, 225
monsoon   7, 8, 11–13, 19, 52, 65, 78, 79, 113, 124, 146, 148, 159, 170, 185, 194, 198, 200, 212, 224, 229, 235, 238, 251
Montgomerie, T.G.   48, 51
Mountain magazine   165, 168
Mount Everest Foundation   221n., 255
mules   177, 183, 184, 198
mushrooms, magic   237

N
Naik Nima Dorje ('N.D.'), Sherpa   229, 230
Namgyal, Rinzin(g) ('R.N.')   55, 57–8, 59–62, 109, 111, 116–117, 121, 122
Nanga Parbat   6, 118n., 173, 175, 190, 197, 198, 199
Nango La   57, 72, 73, 212
Nat(h)u La   205
Nehru, Jawaharlal (Pandit)   207, 209
Nepal
    East(ern) Nepal   vii, 7, 8, 9, 12, 18, 20, 27, 44, 47, 54, 57, 59, 62, 65, 69, 93, 99, 111, 154
    government   43, 181, 184, 216, 221
    Maharajah   183, 184
    territory and history   vii, 6–8, 14, 15, 17–28, 36–8, 42–6, 50–4, 57, 59, 62, 65, 69, 84, 93, 94, 99, 107, 117, 119, 124, 132, 141, 147, 161, 165, 175, 181, 183, 185, 188, 207, 209–12, 216, 221, 222, 225, 229, 236, 248
Nepal Gap   113, 152, 154, 155, 173, 188, 190, 191, 200
Nepal Peak   9, 119, 152, 173, 188, 198, 200
Newari   8, 19, 21, 23, 25, 26–7, 28
Nike, Sherpa   235
Nima Tendrup, Sherpa   184

Nima Tenduk, Sherpa   165
Nima Tsering II, Bhutia   198
North, Marianne   86–7, 89
Noyce, Wilf   208

O
oak   15, 67, 113, 159
Ochterlony, Major-General David   44
oedema   168, 229, 251
oxygen, bottled, use of   35, 91, 158, 221, 222, 223, 227, 228, 229, 233, 234, 246, 250, 254, 255, 255n.

P
Pache, Lieutentant Alexis   135, 139, 141–5, 214, 216, 221, 226, 228
pagoda   23, 24, 25, 234
Pakistan   17
Paidar, Herbert   173, 200, 201, 202, 212
Pallis, Marco   179
Pandim (Pundim)   10, 76, 77, 89, 91, 93, 105, 124, 159, 169, 170
Pangaea   6
Pang Lhabsol   228
Pangpema (Pangperma)   55, 94, 117, 118, 153, 183, 185, 239, 240
Pargatzi, J.   210, 211
Paro   36
Pasang, Sherpa   173–5, 193, 194, 199, 204, 205, 211, 229
Pasang Chakadi, Sherpa   173
Pasang Kikuli, Sherpa   173, 175
Pasang Phuttar, Sherpa   171
Pashupatinath temple   22
Pathan   20, 22, 23, 24, 25
Patna (Palaliputra)   17, 18, 36
Pauhunri   80, 151, 153, 154, 155, 156, 208
Pemayangtse   28, 51, 54, 58, 77, 104, 122, 165, 216
Pemba Dorje, Sherpa   228
Pertemba, Sherpa   234
Phalut   84, 123, 140, 216, 225
Phoktanglungma   *see* Jannu
Photographic Society of India   163
photography   83, 85, 95–99, 104–5, 110–11, 114, 119, 139, 151, 156, 163–4, 165, 166, 168, 181, 190, 198, 215, 222
pines   15, 67, 70, 78, 159, 212
    blue pine   15
    chir pine   15, 67
    silver pine   15
pipal   17, 17n., 236
Pircher, Hans   175, 194, 195
Pokhara   27

Proceedings of the Royal Geographical Society   126, 126n., 130n.
pundits   21, 51, 51n., 52, 54–8, 62, 116, 132
Puranas   5
Pyramid Peak (Pathibhara)   9, 119, 127, 179, 211

R

Raeburn, Harold   151, 159–162, 163, 166–7, 176, 214
Rai   22, 27, 28
Ramthang (place and glacier)   10, 55, 117, 118, 187, 210, 239
Ramthang Peak   10, 55, 187, 240
Rangit (Rangeet), Great   11, 49, 74, 75, 76, 78
Rathong La   214, 216
Rathong Peak   9, 74, 141, 162, 171
Renzing, Sherpa   177, 178
Reymond, C.A.   135, 139, 142–5, 221
rhododendron   15, 52, 54, 69, 78, 80, 113, 156, 159, 164, 171, 175, 238
rhubarb (wild/giant)   15, 114, 153
Richter, Helmut   181
Rigo de Righi, Alcesti C.   135, 139, 142–5, 153
Rig Veda   5, 5n.
Roberts, Jimmy   128
Roerich, Nicholas   viii, 90–1, 92
Rothong   104
Royal Botanic Gardens   83, 86, 87
Royal Geographical Society   48, 123, 126, 127, 130, 158n., 163, 187, 215, 221
Royal Photographic Society   163
Rubenson, Carl   126, 128, 130, 131, 147, 149, 170

S

sal (also 'Saul')tree   78, 104n.
Salameh   221
Schäfer, Dr Ernst   177–8, 202–5
Schaller, Hermann   174, 194, 229
Schlagintweit, Adolf   84, 123
Schlagintweit, Hermann   83, 84–5, 123
Schlagintweit, Robert   84, 123
Schmaderer, Ludwig   173, 200–2, 212
Schneider, Erwin   181–2, 184–5, 187–8, 194, 200
Schoberth, Gustav   169–72
Scott, Martha   233, 233n., 236
Scott, Rosie   236, 236n.
Sebu La   80
Sella, Erminio   99, 111, 115, 117, 123
Sella, Vittorio   98–99, 110, 114, 118, 123, 139
Sentinel Peak   155
Shebbeare, E.O.   190
Sherpanis   237, 238, 239, 240
Sherpas   *see under individual names*

Sherwill, Major James   50, 83, 103–6
Sherwill, Captain W.S.   50, 103
Shigatse (Xigatse)   36, 38, 57, 205
Side, Douglas   107, 221, 221n.
Sikkim
   chogyal   29, 51, 69, 75, 76, 81, 82, 104, 203, 207, 209, 224, 228
   dewan (prime minister)   76, 78, 79, 82, 141, 209, 224
   government   49, 65, 82, 84, 111, 171, 209, 224
   history   18, 20, 28, 40, 42, 44, 49, 57, 82, 97, 104, 121, 228
   territory and culture   vii, viii, 7, 8, 9, 11, 12, 14, 18, 27, 28, 29, 30, 39, 50–5, 58, 59, 61, 62, 64, 65, 69, 72, 74, 76–9, 81, 83, 84, 89, 91, 92, 96, 97, 103, 104, 108, 109, 111, 112, 115, 117, 119, 124, 131, 147, 148, 151–5, 160, 162, 163, 169, 170, 174, 176, 179, 182, 183, 190, 198, 199–204, 207–9, 211, 212, 215, 224, 225, 244, 246
Simla (Shimla)   8, 11, 96, 228
Simvu (mountain, pass, col)   10, 152, 159, 175, 178, 179, 191, 193, 195, 199
Singalila Ridge   vii, 8, 9, 10, 27, 49, 64, 74, 84, 92, 103, 107, 123, 141, 169, 188, 225
Singh, Sukhwinder   229
Siniolchu(m)   10, 92, 97, 99, 114, 173, 191, 193, 195, 198, 199, 200, 244
Siwaliks   7
'Smile'   237–8
Smythe, Frank   10, 13, 108, 170, 182–5, 187, 188, 213, 240, 241
Smythe, Tony   182
snow cave (*see also* ice-cave)   244, 245, 249, 251
Somervell, Howard   91–3, 94
Sona, Sherpa   153, 155
Sonakhoda   48
Sonamarg   223
Sonam Topgay, Sherpa   165, 166, 168
South Langpo glacier   153, 154
Spencer Chapman, Freddy   173, 179, 199, 211
Sphinx, the (mountain)   179, 211
Stone Age   17, 20, 236
storms (snow-, sand-)   vii, 99, 107, 112, 113, 125, 153, 154, 159, 175, 176, 179, 188, 194, 211, 243, 245, 250, 251, 253
Streather, Tony   222, 226–8
Survey of India   47, 48, 50, 51, 58, 59, 111, 124, 127, 128, 132, 133
Surveyors General of India   43, 44, 47, 62, 111
Sutter, Alfred   210, 211

INDEX

T

Talung glacier 11, 105, 124, 156, 159, 163–5, 172, 175, 176, 190
Talung Peak 9, 161, 162, 214
Talung Saddle 161, 166, 167n., 216
Talung Valley (and river) 59, 159, 160, 175, 176, 195
Tamur (Tamar) vii, 8, 62, 67, 69, 72, 212, 234, 237
Tang Kongma 211
Tanner, Colonel H.C.B. 52, 56, 58, 59, 61, 62, 128
Tashi Dorje, Sherpa 230
Tashi Lhunpo 36, 52, 57, 117
Tasker, Joe 233–6, 241–6, 248–56
Tendechar ('Tencheddar'), Sherpa 184
Tent Peak (Kirat Chuli) 9, 198, 200, 201
Tenzing Norgay, Sherpa 171, 214, 222
Terai, Nepali 7, 17, 19, 21, 25, 42, 44, 78, 132
Thapa, Mohan Bahadur 234
The La 115, 117
Thoenes, Alexander 190, 192
Tibet
    government in exile 71
    land and frontier viii, 6, 8, 11, 19, 19n., 20, 22, 23, 25–9, 35–8, 36n., 40, 41, 41n., 42, 44, 49, 51–9, 54n., 61, 63, 67, 69, 70, 71, 72, 78–82, 84, 92, 96, 104, 117, 151, 176, 177, 179, 184, 190, 195, 201, 202–5, 207, 209, 212, 239, 240
    language 19, 30, 49n.
    people and culture 23, 28, 29, 49, 50, 55, 58, 60, 69, 70, 71, 73, 76, 81, 112, 117, 176, 184, 202, 203, 204
Tibetan gazelle 15
Tibetan plateau 6, 11, 12, 65, 66, 67, 79, 81, 84
Tibetan wild ass (*kyang*) 204
Tilman, Bill 164, 165, 174, 175–8, 204, 210
Tista vii, 7, 8, 9, 11, 12, 49, 62, 75, 78, 79, 81, 169, 176, 177, 195, 203
Tobin, Lieutenant-Colonel H.W. 159, 160, 168, 176, 182, 184, 190, 208
Tombazi, Nikolas 163–4, 178
Tonglo 64, 66–7
Tongshyong glacier 163, 164, 175, 176, 178
trees *see individual names*
Trisuli 20, 45
Tsangpo, Yarlung 38, 42, 50
Tsaparang 36, 37
Tseram 53, 60, 107, 119, 141, 161, 165, 183, 184, 185, 214, 216, 225
Tsisima glacier 56, 212
Tso (Lake) Lhamo 80, 81, 84, 154
Tucker, John 84, 215, 216
Tuny, Sherpa 153, 155

Twain, Mark 88–9
Twins, the (Gimmigela I and II, and glacier) 9, 118, 152, 173, 174, 185, 191, 193, 198, 199, 200, 240, 241, 244

U

Unsworth, Walt 131, 132
Upper Zemu glacier 113, 114, 174, 191

V

Vedic
    peoples 5
    period 7, 14, 17
    texts 6

W

Waddell, Major Laurence vii, 37, 96, 97, 128
Wessels, Father Cornelius 37, 38
West, John 158
Westmacott, Mike 221, 221n.
White Hall Adventure Centre 233, 233n.
White, John Claude 97, 111
White, Major 140
Wieland, Ulrich 175, 181, 182, 185, 187, 188
Wien, Karl 175, 194, 198, 199, 200
Williams, Alfred of Salisbury 89
Wood Johnson, G.H. 165, 168, 182, 184, 185, 188
Workman, William Hunter 127
Wyss–Dunant, Dr 210, 211, 212

Y

Yalung glacier (and river) 11, 50, 53, 59, 74, 119, 135, 139, 140, 141, 146, 161, 165, 166, 167, 167n., 168, 184, 213, 214, 215, 216, 225
Yalung Kang 9, 10, 59
Yangma (river and village) 54, 57, 72
Yeti 164, 174, 223
Yoksum 54, 184
Young, Geoffrey Winthrop 158, 215n.
Younghusband, Sir Francis 59, 202–3, 204

Z

Zemu Gap 114, 120, 159, 163, 164, 165, 174–7, 191, 193
Zemu glacier 78, 97, 99, 113, 114, 120, 152, 154, 163, 172, 173, 174, 177, 178, 179, 183, 188, 190, 191, 193, 199, 200, 204, 212, 213, 229, 246, 248